T0314215

# Digital Identity, Virtual Borders and Social Media

# Digital Identity, Virtual Borders and Social Media

## A Panacea for Migration Governance?

*Edited by*

Emre Eren Korkmaz

*Oxford Department of International Development, University of Oxford, UK*

Cheltenham, UK • Northampton, MA, USA

Published by
Edward Elgar Publishing Limited
The Lypiatts
15 Lansdown Road
Cheltenham
Glos GL50 2JA
UK

Edward Elgar Publishing, Inc.
William Pratt House
9 Dewey Court
Northampton
Massachusetts 01060
USA

A catalogue record for this book
is available from the British Library

Library of Congress Control Number: 2021932289

This book is available electronically in the **Elgar**online
Political Science and Public Policy subject collection
http://dx.doi.org/10.4337/9781789909159

ISBN 978 1 78990 914 2 (cased)
ISBN 978 1 78990 915 9 (eBook)

Printed and bound by CPI Group (UK) Ltd, Croydon, CR0 4YY

# Contents

# Contributors

**Abdullah Mohammadi** is a researcher contributing to Meraki Labs research work. He is also a 4Mi Coordinator for the Danish Refugee Council (DRC) and Mixed Migration Center (MMC) in Asia Region. He obtained his MA in Demography from the University of Tehran in 2016 and is studying Ethnic and Migration Studies at Linkoping University, Sweden. His main research interests are forced migration, return migration and smuggling networks. Abdullah has extensive field experience in Afghanistan, the Islamic Republic of Iran, India and Germany.

**Aiden Slavin** is an independent policy advisor and technologist working on issues of digital identity, privacy and rights. He is Head of Private Sector Partnerships at ID2020 and Senior Consultant at the Oxford Centre for Technology and Development. He holds an MSc from the University of Oxford and a BA from Columbia University.

**Deniz Yetkin Aker** is Assistant Professor at the Department of Political Science and Public Administration at Namık Kemal University, Turkey. Her research interests are international migration, high skilled migration and citizenship in a comparative manner.

**Emily Savage** is a Co-Founder of Meraki Labs, a consulting firm focused on humanitarian response to migration and displacement. She has worked in humanitarian contexts and conducted research on displacement and migration across the Middle East, the Horn of Africa, East and Southern Africa. She has held senior country humanitarian positions and has consulted for humanitarian and UN agencies. Emily has an MA in Human Geography from McGill University

**Emre Eren Korkmaz** is a political scientist and international relations expert and has been working as an academic at the University of Oxford's Department of International Development since October 2016. He is teaching the MSc in Migration Studies, and was granted Junior Research Fellowship at St Edmund Hall. Emre is also a research associate at the Centre for Technology and Global Affairs of the Department of Politics and International Relations.

**Erin Harris** is a graduate of the MSc in Migration Studies at the University of Oxford, where her current research focuses on new border security technologies and their impact on the US immigration system. Erin's previous professional experience includes working in the field of US immigration law, where she primarily handled asylum cases.

**Johanna Bankston** is a graduate of the MSc in Refugee and Forced Migration Studies programme at the University of Oxford. Her research interests are in digital rights, migration governance, and the potential of technology to facilitate safer passage for migrants. Ms. Bankston has worked as a research assistant for the Oxford Department of International Development and the UC Berkeley Human Rights Center's Digital Investigations Lab. She also worked as an asylum paralegal for more than two years in San Francisco, specializing in LGBTQ+ and gender-based violence claims.

**Margie Cheesman** is a digital anthropologist based at the Oxford Internet Institute. Her PhD research examines the use of blockchain technologies in humanitarianism, and involves fieldwork in Jordan. Margie's research interests include money and identity infrastructures, inequality and rights. She has worked for the civil liberties initiative, Open Migration and the ESRC Centre for Research on Socio-cultural Change (CReSC). She has conducted research with international organisations such as InfoMigrants, GSMA Mobile for Development, and United Nations agencies. Margie is Assistant Editor of the journal *Big Data & Society*, and her editorial work includes *Data Politics: Worlds, Subjects and Rights* (Routledge, 2019) and *Data Practices* (forthcoming).

**Roxana Akhmetova** is a D.Phil. in Migration Studies student at the University of Oxford. Broadly, her research is focused on the ethics behind the use of artificial intelligence and machine learning by Canadian immigration agencies. Roxana has an MSc in Migration Studies from the University of Oxford and an MA from Canada.

**Ruta Nimkar** is a co-founder of Meraki Labs. She was formerly the Regional Head of Programs Central and South West Asia and the Country Director Iran with DRC. She has humanitarian experience in Iraq, Afghanistan, Somalia, South Sudan and Kyrgyzstan, with the Danish Refugee Council, Cesvi, ACTED and IOM. Her main research interests are forced displacement and mixed movement flows, as well as diasporas; sectorally, she works in protection and livelihoods. Ruta has an MA in International Relations from Yale University and a BA in Economics and Management from the University of Oxford.

# 1. Introduction to *Digital Identity, Virtual Borders and Social Media*

**Emre Eren Korkmaz**

This book discusses how states deploy frontier and digital technologies to manage and control migratory movements and how they cooperate with tech corporations in this field. Blockchain, artificial intelligence/machine learning (AI/ML), social media, big data analysis and various smart border solutions offer a wide range of opportunities for states to deal with legal and irregular migrants, refugees and asylum-seekers. However, this is an emerging area, these tech-based projects are still experimental and they are understudied topics in the literature. This book aims to contribute to the literature as authors critically examine the consequences of these technologies and evaluate them from the perspective of rights and liberties of migrants.

Despite the obvious power imbalances, technology remains an arena of political struggle. The same technologies can be used to promote peace and spark conflict. For this reason, the political struggle to make corporations and governments accountable can still change the course of technological development. For instance, if workers are concerned about automation or the use of AI by corporate HR departments, trade unions can raise these issues in the public square and political decision-makers can take action to upgrade skills, guide the automation process and grant unions new rights to monitor the decisions of algorithms. For people worried about automated weapons and military AI algorithms, various organisations devote their energy to controlling or even banning these weapons to maintain the peace. Anxiety over fake news and manipulative campaigns undermining democracy and the trustworthiness of elections is spurring many non-govermental organisations (NGOs) and political movements to advocate for democratic principles and make political parties accountable for their social media advertising. Finally, those worried about tech corporations' ethical missteps and commercial irresponsibility can lobby governments to take the steps necessary to write regulations to ensure accountability and transparency.

However, migrants and refugees do not have the same access to the arena of political contestation to protect and advance their interests. For anyone wishing to observe the likely shape of a dystopian future of a tech-oriented society in which people lose their political autonomy, a close look at the daily experiences of migrants and refugees vis-à-vis the large tech providers will provide sufficient clues. This book aims to challenge such a totalitarian and hegemonic approach to the people on the move and provide cases demonstrating how such technologies are deployed for the interests of states and corporations and deny the basic rights, liberties and agencies of migrants and refugees.

This book consists of six chapters. The first two chapters focus on the blockchain-based digital identity initiatives, two others discuss different forms of smart/virtual borders, comparing the US, Canada and the European Union (EU) which highly invest in AI. The sixth chapter examines the role of social media and mobile applications on smuggling activities and irregular migration and the last chapter evaluates how social media posts against refugees had an impact on the European Parliament elections in 2019.

In the second chapter, Margie Cheesman and Aiden Slavin discuss an important issue on blockchain-based self-sovereign identities (SSI) with the aim of empowering refugees. They analyse the main arguments and approaches of two SSI start-ups and demonstrate how in practice they fall into the trap of their initial critiques towards corporations and states instead of revolutionising the identity management. When I read the chapter, my conclusion was that individualist approaches could not be a genuine alternative to oppose monopoly power of corporations and states. Radical discursive suggestions without any critique of capitalism and ignoring alternative public spheres do not allow these initiatives to keep their promises.

In the third chapter, Aiden Slavin sheds light on the historical background of identification technologies and compares today's digital identity initiatives with the Nansen Passport invented in 1921 to offer a recognisable form of identification to refugees of the Russian civil war. Aiden's case study is the Blockchain Pilot Project in Kenya and he points out similar motivations to identify forcibly displaced populations. Aiden argues that digital identification and the Nansen Passport are similar in so far as both:

- are justified based on the extension of rights and services to the beneficiary
- are the product of entanglement of actors from both the public and private sectors
- make possible control over the individual in order to enable extractive practices on behalf of institutions
- begin as efforts to render communities 'on the periphery' of global power 'legible'.

In the fourth chapter, Roxana Akhmetova and Erin Harris compare the deployment of AI algorithms in the US and Canada for migration governance to control and monitor borders and automate decisions of visa and asylum applications. This chapter demonstrates the risks of automating border security and asylum decisions from the rights of migrants and exposes the role of tech corporations working closely with military, security and intelligence institutions to mine the data of migrants from counter-terrorism and security perspectives.

The US and Canada are at the forefront of incorporating AI and ML technologies into border control and migration governance; however, the extent to which the two governments are incorporating and testing new AI and ML technologies varies and the main argument of the chapter is that the use of AI immigration enforcement technologies infringes upon migrants' rights of privacy and protection while increasing discrimination to diminish the agency. In attempts to create 'smart borders', Roxana and Erin argue that both Canada and the US are using AI technologies to erect 'invisible border walls' that are diminishing government responsibility and accountability for their actions and decisions as well as to refuse entrance to undesirable migrants. Another motivation is the multi-layered interests of private technology and security and arms companies.

In the fifth chapter, Johanna Bankston discusses the virtual-smart borders in the EU and focuses on the use of frontier technologies to counter smuggling activities. Johanna focuses on internet governance at the micro and macro levels and compares competing strategies of migrants, the EU and social media corporations. She argues that EU member states engage in internet governance to deter irregular migration through 'virtual border enforcement' and concludes by emphasising the dangers of creating digital policies and partnerships aimed at subverting human smuggling which would eclipse migrants' digital rights and identities.

In the sixth chapter, Ruta Nimkar, Emily Savage and Abdullah Mohammadi analyse the relationship between smuggling activities and new technologies from the perspectives of smugglers in Afghanistan. Based on their fieldwork results, Ruta, Emily and Abdullah contend that smuggling networks are building an alternative mobility regime and that mobile technology is augmenting the network capital required to build this regime. They use network capital to analyse the demand, supply and organisational aspects of smuggling networks and argue that mobile technology improves the logistics and organisation of smuggling networks, but has not, to date, significantly affected trust between various networks and actors.

In the last chapter, Deniz Yetkin Aker evaluates the role of social media posts that shape the attitude of the European Parliament election voters towards migration. By using Eurobarometer data (91.5) and by focusing on the European Parliament (EP) 2019 Post-Electoral survey, this study compares three EU member states – Finland, Hungary and Bulgaria – where the Facebook posting themes and sentiments of refugees are generally negative. Deniz discusses whether voting for different parties is related to individuals' attitudes towards migration.

The following issues are expanded upon below:

- digital identity initiatives
- a question on the future of voluntary consent when United Nations (UN) agencies and humanitarian NGOs employ frontier technologies in their projects
- how social media and smartphone applications could be converted to resistance tools by migrants and refugees to oppose policies of states and corporations.

## DIGITAL IDENTITY INITIATIVES

Blockchain has often been in the news lately thanks to the speculative cryptocurrency market. It is increasingly being touted as something of a magic bullet in a whole range of disciplines, and has been put forward as an answer to ongoing challenges in the field of international development, humanitarian aid and refugee studies.

I suggest that while blockchain may be used to support people in need, we have to bear in mind that these are not just technical issues – they are political questions. Focusing solely on the technical side is unlikely to bring the promised progress – and might even have the opposite impact.

In recent years, new partnership models bringing together technology companies and humanitarian NGOs have been promoted as a way of supporting refugees and displaced people – for example to facilitate payments and data protection and to support local businesses. A particular focus of such initiatives is the role of distributed ledger technology (DLT) to provide services rapidly and cheaply through cryptographic security without intermediaries, while also offering transparency and accountability.

According to these initiatives, over one billion people have no form of identity, but with the help of technology, multi-stakeholder collaboration might provide these people with digital identities that would allow them to access essential services, open a bank account and receive aid and social benefits. And as it is digital, the technology offers the necessary conditions for accountability and governance to improve the efficiency of the delivery of development aid.

Jim Yong, the World Bank Group President, has described digital identity as 'the greatest poverty killer app we've ever seen'. If such initiatives are successful, according to Mastercard Vice Chairman Walt Macnee,[1] an additional 500 million new 'consumers' and 40 million 'new merchants' can be brought into the global economy from among the world's 2 billion 'unbanked' people. 'The firm believes that greater financial inclusion around the world is a path to long-term sustainable economic growth.' The World Bank Group released a paper on the role of financial services in humanitarian crises in April 2017.[2] The report argues that in order to integrate humanitarian programmes with a developmental approach, refugees and low-income households in host societies should have access to quality financial services, so they can save, invest and receive loans. Donors may also inject liquidity into local financial markets to support market players.

Other voices have been less positive, however. Mercy Corps contributed to the report as follows: 'delivering aid through e-transfers does not automatically lead to the uptake of new financial services by program participants. Instead, participants typically withdraw their full transfer when it becomes available and rarely use their new accounts after programs end. This holds true in both large government social safety net programs and humanitarian cash transfer programs.' This suggests that such proposals to support refugees do not necessarily have approval from the refugees themselves.

Such initiatives are not limited to digital identity. The UN World Food Programme (WFP) has introduced a pilot project called 'Blockchain Against Hunger'[3] deploying technology in Jordan's Azraq camp to make cash-based transfers to 10,000 refugees. WFP relies on biometric registration data provided by the UN High Commissioner for Refugees (UNHCR) and refugees can shop from local supermarkets using iris scans; the system confirms the identity of the refugee in this way, checks their account balance and confirms the purchase.

However, we need to ask some more questions. Technological progress may help us to reach and track any person in need. It may offer them digital identities, mobile numbers and bank accounts. But do all these mean that people are empowered? Does all this progress provide a solution to poverty? And what are the risks?

Firstly, we should not overstate the value of such technologies as emancipatory tools.

In the most positive scenario, this support would help some refugees to set up their own businesses or become waged employees (in countries where they are permitted to work). Thus a Syrian refugee living in one of the neighbourhoods of Istanbul or Amman, for example, would have similar financial and economic opportunities to his or her local neighbours. So they would compete in the same job market, earn similar wages – or have an equal probability of being unemployed.

And in fact the majority of refugees live in the global South and these host countries face serious socio-economic structural problems. Thus refugees would face the same obstacles as locals: high unemployment rates, instances of modern slavery, harsh working conditions, an abundance of cheap labour and constraints on the right to organise and bargain collectively. And even in the countries of the global North, working conditions are increasingly precarious. So even if blockchain technology allows refugees to obtain a digital ID and receive aid or loans from agencies securely, this would not automatically mean that they are empowered or able to overcome poverty.

Secondly, there is a risk of abuse.

Digital identities, bank accounts and mobile phones allow corporations, donors, international agencies and local-national authorities to track people's choices and desires. Such control might allow authorities and corporations to increase surveillance over refugees. An authoritarian state could use such data collected from refugees against refugees – or nations of the global North, which have no sympathy for the movements of refugees and immigrants towards their countries, could use such information to keep refugees in neighbouring countries.

Even in institutionally democratic countries, there are debates about mass manipulation through fake news, interference in democratic elections, the use of search engine algorithms to offer people tailored search results or expose them to different advertisements based on their preferences and choices. Thus millions of people who are not classified as vulnerable can be manipulated and/or mobilised based on the power of corporations and their technological superiority. Therefore, it is also possible that this power could be used to mobilise refugees in a particular direction or discourage them from making certain decisions. Thus they could end up in a more vulnerable and dependent position.

Thirdly, we need to be aware of the motivations of the corporations involved.

As we have seen from the language used – 'underserved customers', 'merchants' – corporations do not define their efforts as philanthropy and we can assume that they expect some profit from their investments. The debate on addressing the problems of 60 million refugees and displaced people may shift to discussions on those 2 billion people without bank accounts who are therefore potential customers. As refugees and displaced people live in other countries for decades, any project towards refugees requires investment in their host countries, and an important portion of the population who are outside formal financial services live in those refugee-hosting countries. Therefore, any investment to formalise refugees will help corporations to reach locals as well. Refugees may be seen as a good starting point for further business goals.

Proposals for using blockchain technology could address some of the basic challenges faced by refugees, displaced and stateless people. However, it seems rash to present such technological progress as a genuinely emancipatory tool.

All such issues – conflicts, climate change and the refugee question – are political questions and they cannot be resolved without political, social and economic solutions; technological answers alone are unlikely to be sufficient.

In addition, it is worth noting that these initiatives seem to offer a top-down solution: a group of people representing giant technology corporations, the UN agencies, humanitarian NGOs and governments, meeting in Munich, Davos or California to empower 'poor immigrants'. But as the Mercy Corps' comment shows, the proposed solutions might not necessarily be accepted by refugees. So, there is a need to listen to and understand the real expectations and desires of people in need and on the move.

## Voluntary Informed Consent (?)

The second concern on deploying these technologies in the humanitarian setting is about the informed consent. In the humanitarian literature, the informed consent of refugees is a widely discussed concern. Informed consent means that signing up to a particular project is voluntary and refugees should have the chance to opt-out. Concrete problems arise when technological solutions are introduced in these projects. Because both refugees and the staff delivering services in the field lack the expertise to assess the ethical and practical implications of these technologies.

The UN agencies are increasingly preferring to employ tech-based solutions in their projects. The main reasons are to be more efficient and obey the 'know your customer' requirements of financial institutions for cash transfers. But, what would happen to people who are not willing to share their data, for any reason, be it political, religious or individual? Such a decision would have a huge impact on the future of voluntary informed consent.

## Technology as a Resistance Tool for Migrants

Nevertheless, despite the vast power imbalance, refugees do benefit from information and communication technologies (ICTs). The flow of information from networks is vital en route and in making choices after arriving in the country of destination. Mobile applications and social media help refugees to connect with the UN agencies and humanitarian organisations and are used to stay in touch with families back home.

Therefore, comparing and contrasting the ways that new technologies are used to surveil migratory movements with how they are used by refugees to resolve the challenges they encounter will provide insights about the impact of technologies. Neglecting this power imbalance leads to an inevitable romanticisation of the relationship between refugees and their smartphones.

It is also necessary to analyse the digital divide within migrant and refugee communities themselves. It is highly skilled, male and literate refugees and migrants that appear to have disproportionate access to smartphones and mobile applications. Thus, gender and socio-economic conditions have an impact on who can benefit from the advantages offered by smartphones

The tech-based power to track people, and the use of algorithms and drones provide immense opportunities for states to manage migratory movements and for corporations to expand their customer base. In stark contrast to such displays against citizens, this power play against refugees can proceed openly, undermining the political agency of these vulnerable people.

## NOTES

1. https://www.forbes.com/sites/ciocentral/2017/08/07/mastercards-big-data-for-good-initiative-data-philanthropy-on-the-front-lines/%234e3ebe5620dc/ (accessed 12 May 2019).
2. https://www.cgap.org/sites/default/files/Forum-The-Role-of-Financial-Services-in-Humanitarian-Crises_1.pdf (accessed 12 May 2019).
3. https://www.wfp.org/news/blockchain-against-hunger-harnessing-technology-support-syrian-refugees (accessed 12 May 2019).

# 2. Self-sovereign identity and forced migration: slippery terms and the refugee data apparatus

**Margie Cheesman and Aiden Slavin**

*Once people own their own identity, then they're less enthralled to their governments and less subject to adverse situations like natural disasters and wars. So, if someone is ejected from their country, if they've already established self-sovereign identity, they can reconstitute their life.*

Joseph Lubin, Founder of blockchain company Consensys, at the US State Department Blockchain Working Group Event. (Coindesk 2017)

*The current approaches of centralized governmental-based identity systems relying on biometrics have serious limitations with regard to both security and privacy ... A more decentralized and self-sovereign identity system ... is not only more flexible and efficient, but can contribute to securing fundamental human rights, especially in countries with unstable governments and fragile institutions.*

Wang and De Filippi (2020) in their paper 'Self-Sovereign Identity in a Globalized World: Credentials-Based Identity Systems as a Driver for Economic Inclusion'.

## INTRODUCTION: DIGITAL IDENTITY AND REFUGEES

The United Nations Sustainable Development Goal (UN SDG) 16.9 mandates legal identity for everyone in the world by 2030, including free birth registrations. Legal identity, typically issued in the form of government-issued passports or national ID cards, is widely understood as a way to provide people with access to rights, protections and services (World Economic Forum 2018). Access to formal identification is crucial for forcibly displaced persons in securing national or refugee status, facilitating legal movement, family reunification and repatriation, and resisting exploitation and exclusion (Manby 2016). Broad interpretations of SDG 16.9 and the global identification agenda suggest that the ability to prove who you are should include access not just to official and legal types of ID but also to digital and experimental ones (Privacy International 2018). Digital identity regimes have multiplied in migration governance; the use of database technologies, social media datasets and biometric interfaces such as iris scans are prominent examples (Latonero et al. 2019).

Systems such as the UN High Commissioner for Refugees' (UNHCR) Biometric Identity Management System (BIMS) and World Vision International's Last Mile Mobile Solution (LMMS) have extended recognition and protection for refugees with new levels of efficiency, enabling useful insights through big data analysis. However, the trend towards 'governing by identity' has raised significant concerns around surveillance, discrimination and exclusion (Lyon 2008). Digital infrastructures designed to recognise and empower subjects are also used to categorise, track and control them (Sim and Cheesman 2020). Datafied approaches to governance tend to invest digital data with absolute veracity; within this, as Lyon points out (2008:508), 'the stories of the self-attesting body' – the ways in which individuals vouch for themselves – are routinely ignored.

'Self-sovereign identity' (SSI) is a new, digital approach to identification which potentially proposes to amplify refugees' own self-attestations. In other words, enhancing individuals' ability to vouch for themselves to access goods and services using technology. Using the cryptographic database technology of blockchain, SSI is a decentralised form of digital identity which is owned and managed by individuals instead of institutions. As demonstrated by the chapter epigraphs, proponents from sectors like humanitarian aid and technology to academia suggest that SSI could solve major problems with contemporary identification. The expectation is that SSI will revolutionise centralised models of migration governance: advocates of SSI have long problematised the monopolising influence of potentially fallible institutions such as states and corporations in identification processes. SSI has a libertarian, individualist flavour: it is commonly defined as a digital identity that a person or entity 'own[s] and control[s], and which cannot be taken away' (Windley 2020). Advocates argue that, specifically in contexts of forced migration, it could provide refugees with a more empowering, secure, rights-protecting and dependable means of recognition across borders. These beliefs are speculative but increasingly widely held and demand closer examination.

In academic literatures, social scientists have adopted a broadly supportive view of SSI. For example, Wang and De Filippi (2020:10) make the assumption that 'in light of the refugee crisis in Europe, and the increasing number of displaced people who lack a formalized form of identification, today – perhaps more than ever – the quest toward self-sovereign identity has become of crucial importance'. Desale (2020) extols the 'infinite potential' of SSI as a way of securing the recognition and societal inclusion of people fleeing hostile states, suggesting that 'our ways of life and standards of living can radically change due to the use of blockchain technology'. Berg et al. (2018:16) argue that SSI fundamentally challenges the domination of traditional identity institutions and will transform refugee integration processes.

Emancipatory claims made about SSI require critical re-evaluation. From a more critical perspective, Zwitter et al. (2020) show that a range of pressing governance questions are unresolved with SSI, while Goodell and Aste (2019:3) apply a human rights approach to decentralised digital identity and find that the conjecture of blockchain proponents, that the technology can meaningfully shift power relations via decentralisation, is unproven. Advocates argue that blockchain can be used to share hashed identifiers and public keys in order to enable transactions without endangering personal information. Blockchain provides the records infrastructure for public keys. The technology has a high security profile because it provides a means of recording secure, encrypted information. Through our analysis of discourses linked to specific case studies, we show that the proposition that SSI will decentralise identity management and give users more control is not only unproven. We argue that the concept of SSI is fundamentally reconfigured by SSI vendors as they adapt what self-sovereignty means for forced migration use cases.

This chapter investigates the latest frontiers of the discussion around self-sovereign identity through a critical evaluation of two refugee SSI start-ups. We analyse resources and evidence collected since 2017 on these cases, as well as material from the wider SSI community including reports, articles, website materials, white papers, presentations and SSI meetup events. We ask: does SSI actually radically re-orientate digital identity management towards the perspectives, needs, autonomy, privacy and personal control of refugees? We investigate the claims made about SSI in order to tease out how concepts of decentralisation and self-sovereignty are rapidly shifting and, furthermore, how these terms might support the political and financial agendas of actors, new and traditional, in the aid and migration management space. We suggest that these discourses facilitate the reintroduction of traditional intermediaries, and the introduction of novel intermediaries into the identification process. We further argue that the potential for profit in these identification schemes validates critiques of the increasing interpenetration of private agendas in refugee contexts.

The following sections situate refugee SSI discourses within the broader conversations taking place in the SSI community that promote decentralised identification systems. We then analyse two case studies of refugee SSI: Taqanu and The Rohingya Project. Firstly, we unpick how these SSI projects formulate the role of traditional intermediaries, including aid agencies and governments, in their work. Secondly, we outline how the positioning of commercial actors poses concerning implications. Finally, we suggest that the invocation of slippery, ambiguous SSI terminology, blockchain, decentralisation and self-sovereignty, plays a discursive function in eliding these new and persistent relations between refugee subjects and powerful intermediaries.

## KEY DEBATES IN THE BROADER SELF-SOVEREIGN IDENTITY MOVEMENT

One of the key debates in SSI is around what identity is in the first place. Identity is an extremely complex and fluid concept embroiled in thorny philosophical, sociological and legal considerations. It is defined in myriad ways, from the properties to which people exhibit a special degree of attachment, ownership or belonging to, still more broadly, 'the sum of the ways in which we prove who we are to the world in relation to other people and institutions' (Good ID 2019). Identity has become even more of an elastic term in the digital age with the rise of computational approaches to complex socio-political issues. For example, technical perspectives see identity as 'any digital representation of a real-world entity that links a number of a person's attributes' (Bostrum 2011: 13). This chapter is specifically concerned with identification as a socio-political discourse and practice. We see digital identity as part of long-standing efforts to not simply represent individuals and populations, but to construct them, making them 'legible' and therefore manageable to institutions (Scott 1998; Taylor and Broeders 2015). In this sense, we show that SSI ultimately proposes to be no different from previous attempts at the management and classification of refugees.

Refugee-related case studies need to be situated within the broader conversations that have sought to establish and promote SSI as a movement. These discussions have largely taken place in Euro-American settings amongst highly technical communities looking to improve how people are known and knowable online. Discussions of self-sovereign digital identity can be traced to the early 1990s (Kibaroglu 2020). Events like the Internet Identity Workshop, ongoing since 2005, have begun to codify the mission of SSI. The concepts these discussions promote are still largely hypothetical, contested, and few concrete SSI case studies exist in practice. Indeed, interpretations of the term 'self-sovereign identity' are as heterogeneous as the movement seeking to realise it. SSI is a flexible and ambiguous concept that continues to evolve even as this chapter is being written. Nonetheless, three interconnected themes tend to arise (Marlinspike 2012; Rannenberg et al. 2015; Windley 2020): autonomy, decentralisation and intrinsicality.

Firstly, rather than continue to 'slough off data like so much dandruff' (Searls 2018), proponents of SSI would create a system where any individual or organisation could own and control the data about them, only releasing information after granting their consent. Enhancing individuals' ability to own and control personal data is an ideal promoted within SSI discourses as well as beyond this specific context (cf. My Data 2020; Own Your Data Foundation 2020). Blockchain-based data ownership proposes to circumvent the contemporary techniques of information capitalism whereby powerful platforms such as Google, Facebook and connected data brokers extract, surveil and profit from people's online traces with minimal friction.

Secondly, SSI advocates question the monopoly that powerful, centralised institutional actors – referred to as intermediaries – like governments and certificate authorities have over assigning and revoking identity. They seek to disintermediate services and shift the locus of control to technology users and 'position [them] between the issuers and claims inspectors' (W3C 2017). They agree that too much reliance on centralised entities, which are prone to corruption, instability and other political concerns, is disempowering to individuals and an undesirable risk. However, there is no clear consensus on how far the decentralisation should go. SSI could ultimately mean that instead of established intermediaries, the 'individual is their own identity provider' (IBM 2019).

Finally, proponents of SSI also assert that the intrinsic, essentially individual nature of identity must be recognised in contemporary technical systems. These discourses tend to frame SSI as a reaction to the 'problem of identity' on the Internet, which they connect to very broad, age-old philosophical debates about personhood. Christopher Allen, author of the

'Laws of Identity', a foundational text among SSI supporters, describes identity as 'that ineffable "I" of self-consciousness, something that is understood worldwide by every person living in every culture' (Allen 2016). Here, Allen's approach to identity draws on Decartes' famous notion of the independent, knowing ego – that is, the contested idea that human beings have inner selves (Baker 2011). Advocates of SSI argue that, over time, the Cartesian Self was fragmented by institutions, primarily governments and corporations. By rendering identities into identity documents, they argue, institutions made quantifiable units of identity from the once-whole Cartesian Self and enabled the localisation of control in power centres (Allen 2016; Marlinspike 2012). The emergence of the Internet and platform capitalism (Srnicek 2017) is seen as spurring the balkanisation of the Cartesian 'I'. Built with an addressing system designed to identify physical endpoints (machines) and not individuals, the Internet was created without what is referred to as an 'identity layer' (Blockchain Bundesverband 2018). The lack of a baked-in identity layer resulted in the further breakdown of identity in the online world. Identity service providers (ISPs) such as Facebook provide identity verification for other websites in the now-dominant federated model of Internet identification. The resultant dissembling of digital bodies into reams of personally identifiable information (PII), they state, reifies online personas as valuable, commodified assets which individuals struggle to manage securely (Allen 2016; Searls 2018).

SSI is increasingly understood as the antidote to the historic and ongoing balkanisation of people's essential, intrinsic identities. It is defined by an attempt to give users control and ownership over personal data and, in doing so, reconstitute and reclaim their very selfhood in the digital age. Within this focus on individual rights, some SSI proponents advocate that individuals should have multiple forms of ID for use in different contexts, while others privilege the idea of a single, holistic, unitary digital identity. In these variegated discourses, the distributed ledger technology blockchain is lauded as the enabling infrastructure for SSI. SSI advocates adopt the increasingly common supposition that blockchain is a transformative type of technology, the inevitable means by which sociotechnical life will be structured securely, permanently and equitably, with technology users at the centre. Through its algorithmic governance and cryptographic consensus techniques, blockchain believers suggest it replaces the need for trust in institutions with trust in code (Swartz 2016). The idea is that by downloading an SSI application on their local device, users can manage their own information in a highly secure manner and decide what is necessary to release to service providers (IBM 2019).

The efforts undertaken by wider SSI advocates to rebalance identification power and provide individuals with more digital privacy, security, control and agency are well meaning. However, they should be connected with a number of critiques. It is worth referencing emergent scholarship about the dangers of combining ideas about the intrinsic nature of/right to identity with ideals about personal data ownership for all individuals. By reducing data to a personal commodity and applying the logics of property to its purchase and sale, Kerry and Morris (2019) argue that we 'double down on a transactional model based on consumer choice'. This can enable platform-based services to continue justifying the extraction of personal information in exchange for their services. Furthermore, the ability to achieve the desired ends of SSI using blockchain is complicated by the processes required to store and secure information. Storing personal information on a blockchain, by definition an immutable record, could make the data forever vulnerable to corruption or theft, particularly by future technologies with enhanced encryption-breaking capacities. On the other hand, storing data off-chain may have the effect of reincorporating traditional intermediaries into the data storage process. This would directly undermine the claim of radical decentralisation of information made by blockchain proponents (Article 19, 2019). We take up similar concerns about intermediary arrangements in SSI projects in refugee contexts; in these arrangements the social risks around identity governance and information security crystallise.

## SSI MEETS GLOBAL GOVERNANCE AND HUMANITARIANISM

SSI has recently garnered more sustained attention in established settings of governance and population management, including, ironically, among traditional intermediaries like states and central banks. 'Enabling frameworks' for SSI have been explored and written about in the Canadian and European contexts (Preukschat and Reed 2019) and examined by governments at the regional level. The Pan-Canadian Trust Framework, for instance, attempts to codify the rules, processes and requirements necessary to implement a government-led SSI system for Canadian citizens (Bouma 2019). The promise of SSI is now also widely discussed in the sectors of migration governance and humanitarianism. For example, reports and white papers of social impact technology and aid organisations have documented the emergence and development of SSI initiatives since 2016 (Caribou Digital 2016; Coppi and Fast 2019; Stevens 2018; WHO et al. 2017). Interest in what self-sovereignty could offer individuals complements the turn towards user-centric innovation and participatory development (Tacchi 2012). It is increasingly supposed that SSI could not only decentralise and strengthen identification systems for national citizens, but also that it could support populations who face challenges negotiating national identity claims, such as refugees.

SSI has come to be linked to forced displacement in that some of its advocates see the refugee as the paradigmatic figure of globalisation that needs to be reached by an independently managed identity (Caribou Digital 2016; Martinson 2018; Sovrin 2018; UNHCR 2018). For more than a century, international organisations have developed and deployed ways of identifying refugees and stateless persons. Historically, these identifications have been paper-based and issued in the form of certificates of identity, travel documents and passports. Refugees cannot always retain identity documents, especially when they are fleeing persecution and violence. This can later make it difficult or impossible to access critical services. Although the relationship that refugees, stateless persons, internally displaced persons and others have to identification has changed little over the last century (in that identification remains capable of conferring both safety and risk on its bearer), the technologies and datafied processes used by issuing agencies are shifting rapidly. Critics of digital humanitarian initiatives suggest they are contributing to an increasingly surveillant, top-down apparatus which extends the vulnerabilities and harms faced by refugees (Latonero et al. 2019; Madianou 2019b). Research has consistently found that robust, informed consent processes and security safeguards are lacking in contemporary refugee identity management systems. As a result, these systems are prone to data breaches and other issues such as technological function creep (ICRC and Privacy International 2018; Kaurin 2019). By privileging the data control, ownership and privacy of refugees, SSI projects aim to tackle these problems.

Refugee SSI projects include Taqanu and The Rohingya Project (TRP) as well as Banqu, TYKN, Gravity, BitNation and others. Through blockchain-enabled digital identity, they seek to facilitate forms of power and agency for people whose protection under national law is in question. We examine Taqanu and TRP as examples of a trend towards the incorporation of potentially radical SSI into established settings and practices of migration governance. Taqanu has been named a finalist in the 2019 Global Social Venture Competition, recognised as a speaker at the UNHCR Global Virtual Summit, and, in July 2019, pitched to central bankers and policymakers at the G20 Summit in Hamburg. *Guardian*-featured TRP is collaborating with international agencies, including with UNHCR on a financial survey of Rohingya populations in Malaysia (Blockchain for Humanity 2019; Guardian 2018). These are therefore notable case studies seeking to intervene in refugee identification practices.

## ANALYSIS

### SSI Solutions to Refugees' Problems?

Both case studies propose to solve issues with the current state of refugee identification but in different ways. Taqanu's intended users are currently global refugees rather than a specific population. Taqanu seeks to address the generic problems it associates with centralised refugee identification management: in particular, the challenges that displaced, paperless people encounter in obtaining recognition and access to financial services. In comparison, The Rohingya Project (TRP) is rooted in a specific context. It looks to tackle the systemic lack of protection in law, policy and practice that Rohingya people have faced in Myanmar and in other states such as Bangladesh, India and Malaysia. In these states, Rohingya children are denied birth registration (de Chickera 2018). The struggle of Rohingya people is attributed to their lack of legal recognition as an ethnic group, and various attempts have been made to garner recognition for the Rohingya through identification.

SSI projects envision refugees using mobile applications to manage their self-sovereign IDs. Through a digital wallet installed on a local device, users could access a repository of identity claims in order to use social and financial services. For example, TRP has refugees download an 'R-ID wallet', which they can use to store 'geographic, social, linguistic, cultural, and professional' claims such as educational certificates, land titles, and a 'video testimonial' of them. In this example, TRP verifies each of these claims and issues refugees with a cryptography-based unique ID number and barcode. Once verified, refugees could use this to attest to their eligibility for certain services, assuming the providers recognise their app's validity. By these means, TRP intends to give refugees an alternative way of accessing healthcare, education, microfinances, e-voting and other services, outside the mechanisms of any government.

Taqanu also has refugees and people without a fixed address use their mobile phones to build a self-sovereign digital identity. However, they emphasise how social media data can be used in the place of other documentation to make refugees 'bankable'. Taqanu positions itself as an alternative bank whereby individuals 'have their social media data compiled and analysed in a way that makes it possible for regulated banks to verify the refugees' identities to a sufficient degree of probability to offer them basic banking services – even if they can't provide a government-issued birth certificate, passport, or other recognized national identity document' (Chohan 2016). The cases seem to offer a promising model of migration management in which conceptualisations of sovereignty are fundamentally shifted from central authorities to refugees. However, closer inspection of these two case studies reveals a more complicated picture.

## SSI Meets Traditional Migration Management Intermediaries

This section examines the proposed governance structures of our two case studies in order to explicate how both projects are incorporating traditional intermediaries into their implementation models. The projects invoke SSI terminology like decentralisation and self-sovereignty, but alter their meaning in practice. We therefore suggest refugee SSI discourses are ambiguous and misleading.

### Taqanu

In their blog 'Is a Person Without an Identity Still a Person?', Taqanu echoes many of the assumptions raised by proponents of SSI. Chief among these assumptions is the belief that identification practices are the essential basis of personhood – rather than political and bureaucratic systems that *construct* particular forms of personhood towards particular ends. The blog divides identity into three categories: moral, legal and financial. Focusing on the third, Taqanu laments the delegitimisation of refugees as economic actors. Taqanu attempts to bypass traditional institutions – financial service providers, central banks and regulatory agencies – in order to set up economic identities for refugees (Taqanu 2016). They are motivated by solving the challenge refugees routinely face when/in gaining entrance into traditional banking institutions due to the high burden of proof placed on due diligence requirements by Know Your Customer (KYC) and Anti-Money Laundering rules.

Introduced in the aftermath of 9/11, these rules define and mandate identification processes (among other security protocols) for financial service providers. In order to circumvent these requirements, they plan to leverage a 'digital footprint and a person's phone to identify and authenticate people with a very high degree of accuracy and use this newly created digital ID to onboard people to a banking solution' (Taqanu 2016). To meet regulatory requirements and enable access to non-traditional financial services, Taqanu would establish a mobile-only branch-less bank for refugees. In short, Taqanu seeks to enable financial inclusion by providing a self-sovereign digital ID that would give refugee 'customers' ownership of personal and transactional data.

However, these plans are in tension with other ideas that Taqanu has espoused as they enter into more concrete discussions with traditional aid industry actors. Despite the claims to develop user-owned disintermediated identification for refugees that is enabled by decentralised blockchain technology, Taqanu plans to reintroduce old intermediaries into refugee identification management. At the UNHCR Global Virtual Summit on Digital Identity (May 2019), Taqanu advocated for a centralised institutional approach to identification management that would establish the UNHCR as the core authority or trust provider (UNHCR 2018). Referencing the UNHCR's privacy policies and experience with refugee identification management, Taqanu proposed the creation of self-sovereign refugee identity verification – but, paradoxically, this would be managed by the UNHCR. Taqanu opted for an incremental approach to the development of SSI. This may ultimately better protect their intended users, for unlike the technical vendors mentioned, UNHCR has decades of experience and a global mandate to protect refugee communities. However, positioning the pre-eminent global intergovernmental institution of refugee management as the central intermediary shifts the meaning of SSI for the refugee context.

While UNHCR is in many cases the pre-eminent authority of refugee administration, it is ultimately bound to state mandates. Under Articles 27 and 28 of the 1951 Convention Relating to the Status of Refugees, all contracting states are required to issue documentary proof of identity to refugees lawfully residing in their territory. This can be done either in the form of identity papers or valid travel documents. While UNHCR can facilitate the creation of identity documents, it depends on the state for its authority and legitimacy. This relationship makes it unlikely that any SSI initiative in which the UNHCR is involved will ever be fully decentralised. In the past, UNHCR has signed data sharing agreements with governments, enabling direct access to sensitive information (Madianou 2019a). By incorporating UNHCR into its governance, the project connects with and facilitates some degree of government control, rather than obviating or circumventing established patterns of humanitarian information management and refugees' legibility to states.

This departure from the radical dreams of disintermediation espoused by some SSI proponents is not aberrant for Taqanu but endemic: other examples of their written outputs highlight the central role of governments in the implementation of SSI systems (Chohan 2016). Taqanu's mobilisation of the term SSI to new ends exemplifies the shifting associations of blockchain and decentralisation. Taqanu only potentially rethinks the role of the banking sector. It is not obvious how this alternative financial identity would be recognised by necessary parties, for example, when refugees want to pay rent or access a loan or a mortgage. Financial identity is unlikely to help refugees access the most meaningful rights that come with legal recognition and citizenship, like social protection or welfare benefits. The path to achieving digital identity that disintermediates the authorities of migration governance and is meaningfully owned and administered by refugees is also unclear in this case. It is possible that the pressures of providing proof-of-concept and securing investment may force SSI start-ups to cooperate with traditional intermediaries in order to gain access to pilot populations. It is further possible that the needs of implementation in real-world scenarios, particularly refugee contexts, instigate a reconsideration of the radical dreams espoused by wider SSI advocates. These needs underscore the importance, expertise and inescapable authority of historic institutions like UNHCR and international governments.

## The Rohingya Project

Our second case study, TRP, describes itself as a grassroots initiative attempting to combat the socio-political and financial exclusion of Rohingya people in Bangladesh and abroad. On their website, TRP suggests that Rohingya people share 'all of the features of a common ancestry, language, society, culture, and even Burmese nationality (before 1982)'. They highlight that the Rohingya ethnicity has been systematically excluded from recognition in Burma since the passing of the 1982 Law of Citizenship. For many Rohingya, identification is seen as linked to genocidal violence; many have resisted identity registration processes due to concerns of visibility and the risks of persecution that it could entail. Individuals in the Bangladesh security apparatus, for instance, understand identification as key to enabling the eventual removal of the population (Brinham 2019).

TRP positions itself as a reactionary movement tasked with securing the recognition of the Rohingya ethnicity. Rather than entrust centralised financial or public institutions with identification, TRP would create an SSI system whereby identity holders manage how they share their information in unmediated interactions. TRP asserts that it is motivated by an ethos of radical decentralisation. As its founder Mohammed Noor put it: 'Why does a centralised entity like a bank or government own my identity? Who are they to say if I am who I am?' (cited in the Guardian 2018). Echoing broader SSI discourses promoting identification as the essential root of personhood and mainstream social life, TRP aims to offer a way for this stateless group to 'rebuild themselves as a people'. The project seeks to eventually reach the entire diaspora. It plans to begin by registering 1,000 Rohingya in Bangladesh with their version of SSI in order to promote access to a variety of services, such as microfinances (TRP 2020). TRP makes a promising proposal with SSI: to harness the privacy and security benefits of blockchain to enable safe access to recognition, as well as a variety of services including microfinancing, crowdfunding and other forms of financial services (TRP 2020).

One of the key problems with TRP is that it neither confronts how it reflects identity nor how it creates identity, yet both actions have the power to determine the highly sensitive question of who is Rohingyan. TRP uses the language of disintermediation and decentralisation, but positions itself as a new kind of intermediary. In a report assembled for TRP by researchers at the Jackson School of International Studies at the University of Washington, blockchain is touted for its ability to create an acephalous governance structure in which individual identity holders determine what information they share and with whom. Nowhere does it acknowledge the central role that TRP grants itself in determining ethnicity and thereby granting/acting as the gatekeepers of access to the platform itself. Admission to the TRP platform is determined by membership of the Rohingya ethnicity. Determining ethnicity is a notoriously sensitive process rife with potential for abuse, harm and exclusion (Mamdani 2010). TRP uses a multi-part interview to determine membership of the Rohingya ethnicity, focusing on factors like knowledge of cultural history and geography (TRP Webpage). Needless to say, the definition of Rohingya ethnicity advanced and verified by TRP is subjective. Admission to its digital identification system depends on a performative adherence to its understanding of the Rohingya ethnicity. Therefore,

TRP positions itself as a new and unavoidable authority in its project to advance the ethno-centric identification regime. Ultimately, TRP's definition of Rohingya ethnicity, no matter in what way or to what degree it attempts precision, will result in the exclusion of individuals who do not circumscribe to their political imaginaries of Rohingya ethnicity.

TRP is also connected with traditional migration management intermediaries like UNHCR, which it identifies as a partner on its homepage (TRP 2020). This relationship brings TRP's claims to decentralisation and self-sovereignty into question. As discussed, UNHCR is the expert institution in refugee protection. However, it also has a long and complex history with the Rohingya and has contributed to inducing repatriation campaigns in previous circumstances of exodus (Crisp 2018). As with Taqanu, the potential association with UNHCR could involve data sharing with governments. There is uncertainty around what kind of alternative digital service TRP would be and what rights they could provide. A report commissioned in partnership with TRP notes that an agreement among ASEAN member states notes that digital ID can be recognised as a legally valid form for ID, potentially indicating that official governmental institutions are incorporated into the project's plans (Curran et al. 2018). This high-stakes case study has also come under criticism for the devastating risk potential involved in storing personal information about this persecuted group on a blockchain (Piore 2019).

While the project team exhibits an informed and good-willed effort to address a pressing issue, the paradox and power dynamic here – that self-sovereignty still involves TRP giving people a definition of their own ethnicity – remains a fundamental issue. This issue is neglected in discussions around SSI because the language of self-sovereignty and decentralisation misleadingly indicate that it is about circumventing institutions and refugees attesting for themselves. Defining conditions of belonging is a deeply complex and dynamic process. Self-ascribing ethnic or national membership is an extremely political and contentious action, particularly for Rohingya individuals who have faced persecution repeatedly on the basis of their perceived belonging. In these contexts, it is no surprise that technology vendors have shifted away from radical visions of self-attested SSI-style identity and towards the reincorporation of traditional institutions that continue to issue, maintain and revoke identifications. What this means for the recognition of the Rohingya is unclear. But if recent history offers any indication, TRP – like UNHCR – may face significant challenges in realising their own mandate as an intermediary of refugee protection (Crisp 2018).

Taqanu and TRP both invoke blockchain to undergird claims about bypassing retrograde problems of governance and bringing about SSI. For example, TRP references 'zero-knowledge proofs', a cryptographic and unproven means of affirming the validity of an identity claim without revealing any specific information. Although SSI and blockchain are associated with disintermediation and user-centricity, the meaning of these terms has become ambiguous. Traditional intermediary relations and logics of centralisation still prevail in our indicative case studies. This suggests that although they question the long-standing socio-political and economic structures in migration management, they are ultimately unable to circumvent them.

## Refugees, Corporations and the Extractive Turn in SSI

In their new interpretations of self-sovereignty, Taqanu and TRP position themselves as key intermediaries in the identification process and create the potential for specific actors to benefit financially from their role in SSI. The proliferation of connected devices and digital transactions has meant the global market for identity authentication, the process of verifying that someone is who they say they are, is projected to expand from $12 billion in 2018 to $28 billion in 2023 (Banerjee et al. 2019). As with other assets, personal data have become the currency of capitalist life on the Internet (Sadowski 2019). Proponents of SSI have made significant efforts to counter the status quo, conceptualising SSI as the antidote to privacy invasions, data leakage and monetisation, and the concentrations of control over personal data among a small number of providers. However, the term SSI is now also used by projects with extractive business models.

### Taqanu

Taqanu positions traditional intermediaries, including governments, to benefit financially from intermediary status. At the UNHCR Global Virtual Summit on Digital Identity, Taqanu proposed the implementation of an SSI system 'flexible for future developments'. They noted that UNHCR partners and other vendors ought to be able to access different parts of the system and, ultimately, 'reach into that data', with permission from UNHCR. Like all assets, personal data can confer profit on its processor. Taqanu suggests that, through financial inclusion efforts, they will generate 'a store and use of wealth that would otherwise be untapped if portions of the population remained unbanked'. Governments, Taqanu argues, could benefit financially from inclusive policies made possible by SSI. The digital financial inclusion agenda addresses the issue that refugees are systematically excluded from key formal economic rights and activities in most host states. However, it is not simply a well-meaning democratisation of money services to refugees. It is also about enabling private actors in creating and profiting from new markets. Critics have observed that this involves understanding poverty, precarity and disaster as the next frontier for market-making, value accumulation and surveillance (Klein 2007; Roy 2012; Soederberg 2013). Taqanu also plans to leverage the information it harvests as an alternative to traditional forms of credit scoring in order to facilitate refugees' access to financial service providers (Peters 2017). Though Taqanu has yet to release details of plans to provide refugee credit scores to financial service providers, it is possible that, as with most credit scoring agencies, the company could benefit financially from this position. Though unrealised, Taqanu's vision for identification could drive profit for traditional intermediaries, including governments, financial service providers, and potentially Taqanu itself.

**The Rohingya Project**

We highlight the potential for profit in these identification schemes in order to contribute to critiques of the increasing interpenetration of private agendas in refugee contexts. TRP's business model should also be examined. TRP makes data capture a basic requirement for participation on the platform. 'Phase 2' of TRP will seek to provide members of the platform with access to 'a variety of services: micro-financing, social financing and services' (TRP 2020). As with Taqanu, TRP incorporates financial service providers into their plan. The Chief Operating Officer (COO) of TRP is also the Chief Executive Officer (CEO) and co-founder of Ata Plus, a crowdfunding platform operating in the region (Ata Plus Homepage). Their initial call for funding was a joint release among the newly formed Rohingya Project and Ata Plus (Ata Plus Malaysia Facebook Page). The connection between TRP and Ata Plus is unclear, but it is worth following this case as the involvement of the two may indicate the possibility of profit-generation through this SSI system.

Overall, it is evident that refugee SSI initiatives are not exempt from the financial motives of the ever-expanding digital identity industry. These case studies recalibrate the original orientation of SSI, which was towards circumventing the financial value extraction of technology users' identity information by intermediaries. Instead, private actors are positioned to profit from the data flows.

## CONCLUSION: 'SELF-SOVEREIGNTY', 'DECENTRALISATION' AND 'BLOCKCHAIN' AS SLIPPERY TERMS

When deployed in refugee contexts, the term self-sovereignty prompts us to imagine a radical shift in how belonging, asylum and citizenship are formulated. On the surface, it seems to be about individuals controlling identification rather than sovereign states and institutions. The term links back to the SSI movement, which has also promoted alternative solutions to the extractive business models of the contemporary digital identity industry and wider data economy. SSI is located in discourses promoting blockchain as the path to decentralisation and disintermediation. While there are multiple possible types of blockchain, which are not all decentralising, radical and privacy-enhancing (Walch 2017), it is a keyword associated with libertarian agency for individuals and the circumvention of powerful institutional structures. We have found that refugee SSI projects are motivated and enchanted by these ideas. They invoke them when extolling the emancipatory future possibilities of empowering and secure blockchain-enabled digital identity.

However, the invocations of the terms self-sovereignty, decentralisation and blockchain by Taqanu and TRP play a discursive role in eliding centralised and extractive dynamics in their projects. In practice, these terms are slippery and vague. The case studies we have reviewed propose to implement an alternative form of self-sovereignty, shifting the radical principles of the SSI movement in favour of intermediation and new avenues of potential profit. The incorporation of SSI into existing logics and patterns of aid and migration management tells us about the forces at play in the industry. The case studies demonstrate the inescapability of traditional structures of authority such as nation states and UN agencies, the unavoidable political power that defining and constructing identity – ethnic or otherwise – involves, and the market logics which now pervade the public sector. Taqanu and TRP obscure these persistent dynamics because they still accentuate concepts of user-centricity, disintermediation and the intrinsic individuality of identity. In fact, it is not proven or guaranteed that these projects will deliver to refugees meaningful and progressive data management rights, legal status, protection from privacy invasions by governments or extractive data monetisation strategies by companies.

Refugee SSI discourses are shifting rapidly over time. Swartz (2016:101) suggests that radical 'blockchain dreams' have a lot to learn from the less ambitious 'incorporative' ones because they are more realistic. However, as the cases we have discussed show, refugee SSI is being incorporated into migration governance in a way that potentially distracts from new and perennial power dynamics and extractive logics. This can be seen as part of the trend in humanitarianism towards using digital technology in ways that 'lock in' the negative consequences of capitalism and extend patterns of top-down control (Duffield 2016). This also accords with wider accounts of the neoliberal 'co-optation of social justice positions' by institutions in other sectors such as health (Birn et al. 2016). In refugee identification, there is a real risk that SSI efforts will conceal unequal relations among aid industry stakeholders. For refugees and organisations working with SSI providers, this could entail misalignments in understanding and problematic unintended consequences. Extractions of sensitive data and new modes of surveillance and control may be inevitable.

Efforts to reimagine the future of refugee identification should approach SSI and the associated terms with caution. The slippery discourse around blockchain and decentralisation – both radical, utopian and incorporative – should be replaced with close examinations of the root causes of the problems that specific refugee communities face with identification. Technology cannot solve the systemic political, social and economic exclusion of refugees. SSI discourses are now embedded in specific bureaucracies, but do not meaningfully address actual issues with those bureaucratic and political practices. It would be more apt to ask: how could advocacy efforts support UNHCR and other refugee agencies in becoming more socially accountable and privacy-oriented? Or promote pathways to the inclusion of refugees in the established financial sector? Or push for the official recognition, citizenship rights and safety of Rohingya people?

## BIBLIOGRAPHY

Allen, C. 2016. The Path to Self-sovereign Identity. http://www.lifewithalacrity.com/2016/04/the-path-to-self-sovereign-identity.html

Article 19. 2019. Blockchain and Freedom of Expression. https://www.article19.org/wp-content/uploads/2019/07/Blockchain-and-FOE-v4.pdf

Baker, L. 2011. Beyond the Cartesian Self. *Phenomenology and Mind*, 1, 60–69. https://doi.org/10.13128/Phe_Mi-19643

Banerjee, S., T. Misra, S. Sen Kohli, M. Marcus, M. Coden and G. Curtis. 2019. A Great Digital Identity Solution Is One You Can't See. Boston Consulting Group. https://www.bcg.com/en-gb/publications/2019/digital-identity-solution-one-you-cannot-see

Bennet, J. 2017. Myanmar: Rohingya Refugees' Future Unclear as Bangladesh Registers Flood of Arrivals. https://www.abc.net.au/news/2017-09-26/rights-of-rohingya-in-question-bangladesh-myanmar/8987158

Berg, A., C. Berg, S. Davidson and J. Potts. 2018. The Institutional Economics of Identity (25 May). https://ssrn.com/abstract=3072823 or http://dx.doi.org/10.2139/ssrn.3072823

Birn, A.E., L. Nervi and E. Siqueira. 2016. Neoliberalism Redux: The Global Health Policy Agenda and the Politics of Cooptation in Latin America and Beyond. *Development and Change*, 47(4). https://doi.org/10.1111/dech.12247

BitcoinKE. 2020. Uganda Communications Commission to Pilot Blockchain for SIM Registration and Verification. https://bitcoinke.io/2020/07/ucc-to-pilot-sim-card-blockchain/

Blockchain Bundesverband. 2018. Self-sovereign Identity. https://jolocom.io/wp-content/uploads/2018/10/Self-sovereign-Identity-_-Blockchain-Bundesverband-2018.pdf

Blockchain for Humanity. 2019. The Rohingya Project. https://www.b4h.world/project/rohingya-project

Bostrum, N. 2011. The Future of Identity. Nick Bostrum website. https://www.nickbostrom.com/views/identity.pdf

Bouma, T. 2019. Overview of the Proposed Pan Canadian Trust Framework. SSI Meetup page, LinkedIn. https://www.slideshare.net/SSIMeetup/overview-of-the-proposed-pancanadian-trust-framework-for-ssi-tim-bouma

Brinham, N. 2019. When Identity Documents and Registration Produce Exclusion: Lessons from Rohingya Experiences in Myanmar. *Middle East Centre blog*, 1–12. https://blogs.lse.ac.uk/mec/2019/05/10/when-identity-documents-and-registration-produce-exclusion-lessons-from-rohingya-experiences-in-myanmar/

Brown, S. 2019. The Emerging Era of Decentralized Identity. http://files.informatandm.com/uploads/2019/3/121AB_-_TH_-_10.20_-_Brown.pdf

Caribou Digital. 2016. Private-Sector Digital Identity in Emerging Markets. https://goodid-production.s3.amazonaws.com/documents/Caribou-Digitial-Omidyar-Network-Private-Sector-Digital-Identity-In-Emerging-Markets.pdf

Chohan, U. 2016. Is a Person Without an Identity Still a Person? https://www.taqanu.com/backblog/2016/11/12/is-a-person-without-an-identity-still-a-person#:~:text=It%20aims%20to%20help%20humans,the%20work%20that%20Taqanu%20does

Coindesk. 2017. US State Department Seeks Blockchain Boost. https://www.coindesk.com/us-state-department-seeks-blockchain-boost-amid-10-billion-reboot

Coleman, R. 2004. *Reclaiming the Streets: Surveillance, Social Control and the City*. Cullompton: Willan Publishing.

Coppi, G. and L. Fast. 2019. Blockchain and Distributed Ledger Technologies in the Humanitarian Sector. http://doi.wiley.com/10.1111/disa.12333

Crisp, J. 2018. The Untold Story of UNHCR's Historical Engagement with Rohingya Refugees. Overseas Development Institute website. https://odihpn.org/magazine/primitive-people-the-untold-story-of-unhcrs-historical-engagement-with-rohingya-refugees/

Curran, S., A. Kane, S. Anderson et al. 2018. Identities for Opportunities: A Feasibility Study for Overcoming the Rohingya's Statelessness Challenges Via Blockchain-based Digital Solutions. https://jsis.washington.edu/news/identities-for-opportunities/

de Chikera, A. 2018. Statelessness and Identity in the Rohingya Refugee Crisis. Humanitarian Exchange website. https://odihpn.org/magazine/statelessness-identity-rohingya-refugee-crisis/

Desale, A. 2020. Diversity and Inclusion: Blockchain Technology and Digital Identity for Stateless Rohingya Refugees. Kaldor Centre for International Refugee Law. https://www.kaldorcentre.unsw.edu.au/publication/diversity-and-inclusion-blockchain-technology-and-digital-identity-stateless-rohingya

Donner, J. 2019. The Difference between Identity, Identification, and ID. Caribou Digital. https://www.good-id.org/en/articles/difference-between-digital-identity-identification-and-id/#:~:text=Identity%20is%20a%20concept%3A%20the,to%20other%20people%20and%20institutions

Duffield, M. 2016. The Resilience of the Ruins: Towards a Critique of Digital Humanitarianism. *Resilience*, 4(3), 147–65. https://doi.org/10.1080/21693293.2016.1153772

Dunphy, P. and F.A.P. Petitcolas. 2018. A First Look at Identity Management Schemes on the Blockchain. *IEEE Security & Privacy*, 16(4), 20–29. doi: 10.1109/MSP.2018.3111247

Facebook. 2019. Ata Plus Malaysia. https://www.facebook.com/pg/myataplus/posts/?ref=page_internal

Good ID. 2019. Current Definition of Good Digital Identity. https://www.good-id.org/en/glossary/digital-identity/

Goodell, G. and T. Aste. 2019. A Decentralised Digital Identity Architecture. https://arxiv.org/pdf/1902.08769.pdf. http://indicators.report/targets/16-9/

Guardian. 2018. Rohingya Turn to Blockchain to Solve Identity Crisis. https://www.theguardian.com/world/2018/aug/21/rohingya-turn-to-blockchain-to-solve-identity-crisis

Haddad, E. 2008. *The Refugee in International Society: Between Sovereigns* (Cambridge Studies in International Relations). Cambridge: Cambridge University Press. doi:10.1017/CBO9780511491351

Henry M. Jackson School for International Studies. 2018. Identities for Opportunities: A Feasibility Study for Overcoming the Rohingya's Statelessness Challenges Via Blockchain Digital Solutions. https://jsis.washington.edu/wordpress/wp-content/uploads/2018/08/jsis-arp-rohingya-2018.pdf

IBM. 2019. IBM Verify Credentials. https://docs.info.verify-creds.com/learn/ssi-concepts/

ICRC and Privacy International. 2018. The Humanitarian Metadata Problem: 'Doing No Harm' in the Digital Era. https://privacyinternational.org/report/2509/humanitarian-metadata-problem-doing-no-harm-digital-era

ID4D. 2018. Identification for Development Global Dataset. https://datacatalog.worldbank.org/dataset/identification-development-global-dataset

Kaurin, D. 2019. Data Protection and Digital Agency for Refugees. *World Refugee Council Research Paper*, 12. https://www.cigionline.org/sites/default/files/documents/WRC Research Paper no.12.pdf

Kerry, C. and J. Morris. 2019. Why Data Ownership Is the Wrong Approach to Protecting Privacy. The Brookings Institution. https://www.brookings.edu/blog/techtank/2019/06/26/why-data-ownership-is-the-wrong-approach-to-protecting-privacy/

Kibaroglu, O. 2020. Self-sovereign Digital Identity on the Blockchain: A Discourse Analysis. Research gate. https://www.researchgate.net/publication/341946822_Self_Sovereign_Digital_Identity_on_the_Blockchain_A_Discourse_Analysis

Klein, N. 2007. *The Shock Doctrine: The Rise of Disaster Capitalism*. New York: Picador.

Latonero, M., K. Hiatt, A. Napolitano, G. Clericetti and M. Penagos. 2019. Digital Identity in the Migration & Refugee Context. Data & Society. https://datasociety.net/output/digital-identity-in-the-migration-refugee-context/%0Ahttp://files/22755/Digital Identity in the Migration & Refugee Contex.pdf%0Ahttp://files/22754/digital-identity-in-the-migration-refugee-context.html

Lyon, D. 2008. Biometrics, Identification and Surveillance. *Bioethics*, 22(9), 499–508. https://doi.org/10.1111/j.1467-8519.2008.00697.x

Madianou, M. 2019a. The Biometric Assemblage: Surveillance, Experimentation, Profit, and the Measuring of Refugee Bodies. *Television and New Media*, 20(6), 581–99. https://doi.org/10.1177/1527476419857682

Madianou, M. 2019b. Technocolonialism: Digital Innovation and Data Practices in the Humanitarian Response to Refugee Crises. https://journals.sagepub.com/doi/full/10.1177/2056305119863146

Manby, B. 2016. Identification in the Context of Forced Displacement. The World Bank. https://openknowledge.worldbank.org/handle/10986/24941

Marlinspike, M. 2012. What Is 'Sovereign Source Authority'? The Moxie Tongue. http://www.moxytongue.com/2012/02/what-is-sovereign-source-authority.html

Martinson, M. 2018. Blockchain-based Self-sovereign Identity as a Means of Aid in International Humanitarian Crises. https://www.researchgate.net/publication/329017509_Blockchain-based_Self-Sovereign_Identity_as_a_means_of_aid_in_international_humanitarian_crises

Mulligan, C. 2020. Blockchain and Sustainable Growth about the Author. UN Chronicle. https://www.un.org/en/un-chronicle/blockchain-and-sustainable-growth

My Data. 2020. https://mydata.org/about/

Own Your Data Foundation. 2020. https://ownyourdata.foundation/

Peters, A. 2017. This App Helps Refugees Get Bank Accounts by Giving Them a Digital Identity. https://www.fastcompany.com/40403583/this-app-helps-refugees-get-bank-accounts-by-giving-them-a-digital-identity

Piore, A. 2019. Can Blockchain Finally Give Us the Digital Privacy We Deserve? Newsweek website. https://www.newsweek.com/2019/03/08/can-blockchain-finally-give-us-digital-privacy-we-deserve-1340689.html
Preukschat, A. and D. Reed. 2019. Self-sovereign Identity. Manning Publications. https://livebook.manning.com/book/self-sovereign-identity/chapter-1/v-1/1
Privacy International. 2018. The Sustainable Development Goals, Identity, and Privacy: Does Their Implementation Risk Human Rights? https://privacyinternational.org/long-read/2237/sustainable-development-goals-identity-and-privacy-does-their-implementation-risk
Rannenberg, K., J. Camenisch and A. Sabouri 2015. *Attribute-based Credentials for Trust. Identity in the Information Society*. Berlin: Springer. doi: 10.1007/978-3-319-14439-9
Reed, A. and A. Preukschat. 2020. SSI Scorecard. https://livebook.manning.com/book/self-sovereign-identity/by-drummond-reed-and-alex-preukschat/v-1/1
Roy, A. 2012. Ethical Subjects: Market Rule in an Age of Poverty. *Public Culture* 24(1), 105–8.
Sadowski, J. 2019. When Data Is Capital: Datafication, Accumulation, and Extraction. Big Data & Society. https://doi.org/10.1177/2053951718820549
Schoemaker, E., D. Baslan, B. Pon and N. Dell. 2020. Identity at the Margins: Data Justice and Refugee Experiences with Digital Identity Systems. *Information Technology for Development*. https://doi.org/10.1080/02681102.2020.1785826
Scott, J.C. 1998. *Seeing Like a State: How Certain Schemes to Improve the Human Condition Have Failed.* New Haven and London: Yale University Press.
Searls, D. 2018. Harvard Weblog: Data. https://blogs.harvard.edu/doc/2018/09/18/data/
Sim, K. and M. Cheesman. 2020. What's the Harm in Categorisation? Reflections on the Categorisation Work of Tech 4 Good. *Big Data & Society* blog.
Soederberg , S. 2013. Universalising Financial Inclusion and the Securitisation of Development. *Third World Quarterly*, 34(4), 593–612. doi: 10.1080/01436597.2013 .786285
Sovrin. 2018. The Inevitable Rise of Self-sovereign Identity. https://sovrin.org/wp-content/uploads/2018/03/The-Inevitable-Rise-of-Self-Sovereign-Identity.pdf
Sovrin Foundation. 2018. Sovrin: A Protocol and Token for Self-sovereign Identity and Decentralized Trust. https://sovrin.org/wp-content/uploads/Sovrin-Protocol-and-Token-White-Paper.pdf
Srnicek, N. 2017. *Platform Capitalism.* Cambridge: Polity Press.
Stevens, L. 2018. Self-sovereign Identities for Scaling Up Cash Transfer Projects. *Netherlands Red Cross 510 TUDelft.*
Stokkink, Q. and J. Pouwelse. 2018. Deployment of a Blockchain-based Self-sovereign Identity. https://sovrin.org/
Swartz, L. 2016. Blockchain Dreams: Imagining Economic Alternatives After Bitcoin. In *Another Economy is Possible: Culture and Economy in a Time of Crisis*, Chapter 4 (pp. 82–105).

Tacchi, J. 2012. Digital Engagement: Voice and Participation in Development. In *Digital Anthropology*.

Taqanu. 2016. Banking Available for All. https://www.taqanu.com/backblog/2016/10/28/taqanu-banking-available-for-all

Taylor, L. and D. Broeders. 2015. In the Name of Development: Power, Profit and the Gatafication of the Global South. *Geoforum*, 64(October), 229–37. https://doi.org/10.1016/j.geoforum.2015.07.002

Thayer, S. 2018. Rohingya Turn to Blockchain to Solve Identity Crisis. https://www.theguardian.com/world/2018/aug/21/rohingya-turn-to-blockchain-to-solve-identity-crisis

TRP. 2020. Home: The Rohingya Project website. https://rohingyaproject.com/

UNHCR. 2018. UNHCR Now Accepting Proposals on Digital Identity. https://www.unhcr.org/blogs/unhcr-accepting-proposals-digital-identity/

W3C. 2017. Verifiable Claims Data Model. https://www.w3.org/TR/2017/WD-verifiable-claims-data-model-20170803/

Walch, A. 2017. Blockchain's Treacherous Vocabulary: One More Challenge for the Regulators. *Journal of Internet Law*, 21(2), 9–16.

Wang, F. and P. De Filippi. 2020. Self-sovereign Identity in a Globalized World: Credentials-based Identity Systems as a Driver for Economic Inclusion. *Frontiers in Blockchain.*

WHO, World Bank et al. 2017. Digital Opportunities for Displaced Women, Children and Adolescents #ONTHEMOVE. https://www.who.int/pmnch/knowledge/publications/Knowledge_Brief02.pdf?ua=1

Windley, P. 2020. Relationships and Identity. Windley.com. https://www.windley.com/

World Economic Forum. 2018. White Paper Digital Identity Threshold Digital Identity Revolution Report. http://www3.weforum.org/docs/White_Paper_Digital_Identity_Threshold_Digital_Identity_Revolution_report_2018.pdf

Zwitter, A.J., O.J. Gstrein and E. Yap. 2020. Digital Identity and the Blockchain: Universal Identity Management and the Concept of the 'Self-sovereign' Individual. *Frontiers in Blockchain*, 3(26). https://doi.org/10.3389/fbloc.2020.00026

# 3. Digital identification for the vulnerable: continuities across a century of identification technologies

**Aiden Slavin**

## INTRODUCTION

Though questions of digital privacy, security, and identification have long been at the fore of international concern and debate, they achieved new prominence with the release of the European Union's General Data Protection Regulation (GDPR), which entered into effect May 25, 2018 and sent shockwaves throughout Europe. The GDPR is in many ways a response to emerging trends in digital technologies towards the collection and commoditization of personal data. As Dixon, Irish Data Protection Commissioner put it, 'There's a wave coming toward us that we need to push back against' (Satariano 2018). Much of the concern touches upon the topic of digital identification.

A digital identification is a metasystem of digital identifiers, data representing transactions that occur on, with, or through digital technologies. Aggregated and analysed, digital identifiers can reveal a wide range of information about an individual, including uniquely identifying who they are. The possible uses of digital identification are many and include facilitating access to global financial markets, public health services, and voter registration. The proliferation of digital identification also poses serious risks to privacy and security. Several have suggested measures, following on from the GDPR, to curtail the collection and use of individuals' data (BBC 2017).

Though many have expressed alarm at the potential for misuse of digital identification of citizens, few have noted that forms of digital identification are already being leveraged, to borrow a phrase from

Scott, to make vulnerable populations 'legible' (Scott 1998). (That is, apart from the subjects of the data collection themselves.) In 2015, the UNHCR (2015) reported confronting vehement opposition from refugee leaders in the Dadaab region of Kenya. The refugees in question had sabotaged a UNHCR (2015) communications campaign in protest against the implementation of a new biometric identification system. Their concerns were echoed in the Kakuma Daily News Reflector (2016), a Refugee Free Press Outlet that reported that refugees had refused to enroll in a new biometrics identity management system because they felt it 'could eventually risk their individual security as data and identity [were] being shared with third parties.'

The extension of digital identification to vulnerable populations has received little attention. Apart from official reports of International Humanitarian Organizations and statements by the subjects of data collection themselves, few have considered the implications of digital identification for vulnerable individuals. Indeed, discussions of data, security, and identification often appear to be limited in scope to technologically 'advanced' societies, and to largely discount vulnerable populations in their analyses (Lyon 2007). This limitation is not only realized in discourse but substantiated by law: the GDPR, amongst other regulations, has received criticism for failing to protect vulnerable populations in its enforcement (GDPR 2018). Reports that do address the extension of digital identification to vulnerable populations tend to hyperbole. Some celebrate the technology as a way to give individuals control over their data. Some even suggest that, in the case of refugees, digital identification could obviate the need for camps altogether (Juskalian 2018). Others warn that digital identification could threaten lives (Rahman 2017). Apart from their alarmism, what these reports have in common is a focus on the present.

This chapter attempts to moderate the polarity of discourse present in these reports by grounding an analysis of digital identification in an historical context. It examines two cases: the Blockchain Pilot Project and the Nansen Passport. Launched in May 2018 in Isiolo County, Kenya, the Blockchain Pilot Project delivered assistance to more than 2,000 households. As part of the cash transfer program, the Project extended a form of digital identification to vulnerable individuals in drought-stricken Isiolo County. Invented in 1921, the Nansen Passport offered a durable, recognizable form of identification to refugees of the Russian Civil War, enabling them to cross borders and resettle.

Though invented and implemented under radically different circumstances by a host of different actors, this chapter will argue that digital identification and the Nansen Passport are similar in so far as both (1) are justified on the basis of the extension of rights and services to the beneficiary; (2) are the product of an entanglement of actors from both the public and private sectors; (3) make possible control over the individual in order to enable extractive practices on behalf of institutions; (4) begin as efforts to render communities 'on the periphery' of global power 'legible.'

It is one aim of this chapter to demonstrate that, while questions of data, security, and identification often appear new, they have a long history. The Blockchain Pilot Project and the Nansen Passport represent contemporary and historical answers to the fundamental questions of identification: how can one prove that they are who they say they are? Separated by roughly a century, the two case studies demonstrate a series of continuities that reveal commonalities across efforts to extend regimes of identification to the world's vulnerable.

The chapter begins by examining the technology underlying the Blockchain Pilot Project, a cash transfer program. It argues that, through cash transfer programs, a form of digital identification is being created for vulnerable individuals. It goes on to suggest that the contemporary formation of digital identification for vulnerable individuals finds a historical analog in the creation of the Nansen Passport. Both, it argues, sought to provide vulnerable individuals with access to a service or a right. It goes on to demonstrate that the formation of the Nansen Passport was entangled with the formalization of the Modern International Passport.

Though questions of data, security, and identification often appear limited in scope to technologically 'advanced' societies, this chapter makes the argument that the formation of identification technologies for vulnerable individuals is entangled with the extension of identification technologies to non-vulnerable populations (Lyon 2007). I use the term identification technology to refer to a device used to verify identity. Examples of identification technology include passports, drivers' licenses, and national ID cards. I use the term to refer to what others have called identity documents in addition to what I suggest is one of their modern day parallels: digital identification. I discuss a document of identity, the Nansen Passport, in addition to what I argue is one of its modern day paperless equivalents: digital identification.

## BLOCKCHAIN PILOT PROJECT

In May 2018, the International Federation of the Red Cross and Red Crescent Societies (IFRC) in collaboration with the Kenya Red Cross Society (KRCS) conducted an open-loop cash transfer program using a blockchain to record transactions in Isiolo County, Kenya. An answer to calls for drought relief, the Blockchain Pilot Project provided assistance to over 2,000 households.

After securing support from Innovation Norway, a government hub for development, in December 2017, IFRC partnered with private sector partner RedRose to develop a data management platform and construct a blockchain. RedRose offers data management solutions for the Humanitarian Sector. For the Blockchain Pilot Project, its data management platform facilitated consent, recording, distribution, and post-distribution monitoring. The blockchain recorded disbursement on an immutable digital ledger. IFRC-KRCS leveraged the data management platform and blockchain.

Operations began on the ground in May 2018. KRCS, which runs regular needs assessments in-country, determined that Isiolo County required assistance. A frequent aid target, Isiolo County forms one of Kenya's 23 arid and semi-arid lands (ASALs). Drought conditions are prevalent in ASALs, producing high livestock mortality and, in turn, insecurity. Prior to the Pilot Project, KRCS worked with the European Civil Protection and Humanitarian Operations (ECHO) fund to deliver disaster relief to Isiolo County. The Pilot Project targeted the same communities as ECHO, culling beneficiaries from the ECHO registration list. Once complete, IFRC-KRCS uploaded an edited Excel extract containing registration data to the RedRose data management system. Registration complete, IFRC-KRCS initiated beneficiary communications in order to attain the informed consent of its participants.

Beneficiary communications occurred in central locations throughout Isiolo County. IFRC-KRCS made use of a network of local volunteers to communicate a detailed summary of the Pilot and consent statement to beneficiaries in local language. Beneficiary communications addressed a group and then, one by one, secured consent from participants. Their approval was recorded on the RedRose platform.

In order to log disbursement on the blockchain, the Blockchain Pilot Project spawned public keys for beneficiaries. A public key is a pseudonymous representation of a user. In future projects in Isiolo County, the

IFRC has indicated that it will seek to reuse the same public key. Over time information will accrue to the same public key, creating a durable form of identification that could be used across a series of programs.

This chapter argues that the public keys manufactured for the Blockchain Pilot Project constitute a form of de facto digital identification. A digital identification is a metasystem of digital identifiers, or bits and bytes of data that represent transactions. The blockchain constructed by private sector partner RedRose recorded disbursement information while the data management system compiled personal information on vulnerable individuals. While the systems were maintained separately, an interface developed by RedRose enabled IFRC-KRCS to correlate between personal information and transaction data. In order to register for the Pilot, individuals had to have biometrics logged. The Kenya National Identity Card was used as a first-order system of verification. Verification is the process of establishing a correspondence between an individual and an identification. In order to verify identity, the Kenya National Identity Card makes use of biometrics (a thumb print). Thus, the Kenya Red Cross Society and International Federation of the Red Cross issued a form of digital identity to beneficiaries.

## THE NANSEN PASSPORT

Like digital identification, the Nansen Passport also sought to provide vulnerable individuals access to rights and services. In this section, I examine the Nansen Passport as an early instantiation of an identification technology constructed on behalf of vulnerable populations. I go on to argue that the formation of the Nansen Passport was entangled with the formalization of the Modern International Passport, revealing a linkage between the construction of identification technology for vulnerable and non-vulnerable populations.

Demographers estimate that the Russian Revolution and ensuing Civil War sent 1.7 million individuals into flight (Kulischer 1943). Most fled to Poland, the Baltics, and the Ukraine. Among them were some 900,000 of Russian origin whose plight was worsened by Lenin's 1917 denationalization act, which left them stateless and rendered their identity documents moot (Marrus 1985). Their lack of valid documentation frustrated attempts to cross international borders and resettle.

The Nansen Passport entered de facto effect at the International Conference on Russian Refugees, held from July 3 to July 5, 1922. Though it had been previously enacted, few states accepted it as legal tender. In 1922, the Refugee Passport became accepted by 16 governments. In 1924, it was extended to Armenian and, in 1928, Assyrian and Assyro-Chaldean refugees (White 2017).

Several regard the Nansen Passport as revolutionary. As Arendt (1968) notes, the Nansen Passport was the first travel document conferred on stateless individuals. It served as, in effect, the first official recognition of statelessness (Morley 1932). Though at the time, no formal definition of statelessness or refugeehood existed, Skran (1995) argues that the formation of the Nansen Passport can be understood as the origin of Refugee law.

The Nansen Passport was created in order to empower vulnerable individuals, enabling refugees to cross borders, find work, and resettle. The Nansen Passport can be understood as an early identification technology designed to empower vulnerable populations. It was manufactured in an attempt to facilitate the cross-border movement of refugees of the Russian Empire. It was extended to several other groups throughout the 1920s and 1930s in order to empower vulnerable communities.

## CONTINUITIES ACROSS A CENTURY OF IDENTIFICATION TECHNOLOGIES: BLOCKCHAIN PILOT PROJECT AND NANSEN PASSPORT

In this section, I examine four commonalities between the identification technology extended through the Blockchain Pilot Project and the Nansen Passport. I argue that both (1) are justified on the basis of the extension of rights and services to the beneficiary; (2) are the product of an entanglement of actors from both the public and private sectors; (3) facilitate control over the individual in order to enable extractive practices on behalf of institutions; (4) begin as efforts to render communities 'on the periphery' of global power 'legible.'

### 1. Both Are Justified on the Basis of the Extension of Rights and Services to the Beneficiary.

Both the Blockchain Pilot Project and the Nansen Passport were justified on the basis of extending access to rights and services to vulnerable individuals. In May 2018, the IFRC in collaboration with the KRCS conducted an open-loop cash transfer program using a blockchain to record transactions in Isiolo County, Kenya. An answer to calls for drought relief, the program was designed to deliver cash vouchers to beneficiaries in need of an infusion. IFRC-KRCS disbursed cash to beneficiaries through the RedRose platform using Safaricom M-Pesa, a mobile money transfer service with an extensive network in Kenya. Beneficiaries were familiar with the system, facilitating the open-loop cash transfer. Once Safaricom approved disbursement, it sent a record of transactions to the RedRose platform. These transactions were recorded automatically on a blockchain. The blockchain secured the transaction information with cryptography, creating an immutable record. Following disbursement, beneficiaries encashed and spent their funds. Within days of disbursement most had bought foodstuffs, repaid debts, and paid forward education fees. The extension of digital identification was managed on a blockchain, developed with private sector partner RedRose, which IFRC-KRCS argued enabled it to deliver a more secure, transparent, and responsive cash transfer program (CTP), ultimately ensuring that beneficiaries were able to access services (in this case a cash voucher) in a way that was secure and reliable. The extension of digital identification was justified on the basis of the extension of access to rights and services.

Similarly, the extension of the Nansen Passport was justified on the basis of extending access to rights and services to vulnerable individuals. Fritdhjof Nansen—fledgling League of Nations High Commissioner for Refugees—recognized this limitation. He would later note: ‘a great many of the Russian refugees in Europe suffered considerably for the sole reason that they [had] been unprovided with any legal passports or paper of identity, without which they were unable to travel from the countries in which they found themselves’ (League of Nations 1922). To remedy their status, Nansen proposed the creation of an Identity Certificate. Five years after Lenin’s denationalization act, the Certificate was accepted by 16 nations at ‘The Arrangement of 5 July 1922’ (Long 2013). Initially, it was conferred exclusively on ‘personnes d’origines Russes,’ as its frontispiece testified (League of Nations 1922). By 1924, what came to be known as the ‘Refugee Passport’ had been extended to Armenian and later, in 1928, Assyrian and Assyro-Chaldean refugees (White 2017). Nansen justified the extension of the eponymous passport on the basis of extending the right to movement to vulnerable individuals, in this case refugees and stateless persons.

## 2. Both Are the Product of an Entanglement of Actors from both the Public and Private Sectors.

Both the Blockchain Pilot Project and the Nansen Passport were produced by an entanglement of actors from the public and private sectors. In the Blockchain Pilot Project in May 2018, the IFRC and KRCS took a step towards addressing the global identity gap (World Bank 2018). The organizations conducted an open-loop cash transfer program using a blockchain, developed by private sector partner RedRose, to record transactions. By spawning public keys for participants, the blockchain created a form of digital identification for beneficiaries. In future programs in Isiolo County, the KRCs and IFRC will seek to use the same digital identifications. Though it was the Red Cross lending legitimacy to these forms of identification, it was not the Red Cross developing, or even managing, the underlying technologies on which its implementation was dependent. Instead, it was the private sector technology company RedRose. Strapped for resources and strained for human capital, humanitarian organizations are often forced to rely on private sector organizations for the development and maintenance of digital technologies. This is a commonality across several digital identification programs. Another recent program, the World Food Programme's Building Blocks, made use of private sector partners Parity Technologies and Baltic Data Science, an affiliate of Datarella, to build a blockchain for its own program (Sovrin 2018). Thus, the Blockchain Pilot Project was the product of a public-private sectoral approach.

Likewise, the Nansen Passport was the product of an entanglement of actors from the public and private sectors. Private corporations contributed to the formation of the Refugee Passport. Short of repatriation, employment was seen as the optimal solution for the 'problem of persons without nationality.' Many attempts to secure employment through governments, however, failed. States often justified their closed-door policies with reference to the spreading threat of bolshevism and crippled economies (Kerber 2007). States proved unresponsive to calls for employment. In their place, a series of schemes emerged from private corporations, creating jobs for laborers.

These private sector efforts often went against official state policies. Indeed, though the American government had frequently rejected calls to assist refugees, American corporations were eager to secure their labor power. The Blair Syndicate, an organization of American businessmen and elite, secured a loan of one million dollars for the construction of a series of railway lines in Yougo Slavia. The railway scheme proved to be a boon for refugee laborers, who found work after Johnson, Secretary to the High Commissioner for Refugees, secured placement on five lines (League of Nations 1922). This effort found an analog in 'The Serbian Scheme,' a railway built in the eponymous nation that would employ Russian refugees from Sofia, Constantinople, and Athens (League of Nations 1922).

These arrangements, often negotiated directly between the Office and private corporations, epitomize the collaboration between the private sector and non-governmental organizations in the effort to empower refugees through the identity document. This collaboration was not unique to the relief effort but extended to the formation of the Nansen Passport. Private corporations and shipping companies played a role in the formation of the Refugee Passport, participating frequently at the July 3, 1921 Conference when the Refugee Passport entered into effect. Indeed, it was demands for verification from laborers lodged by private corporations, as much as from governments, that gave the Refugee Passport its utility (League of Nations 1922). Thus, private corporations played an active role in the formation of the Refugee Passport. The Nansen Passport thus empowered individuals to cross borders, find work, and resettle.

The involvement of private sector organizations has reopened political, ethical, and legal questions revolving around notions of informed consent, intellectual property, and experimentation. At the center of these old concerns is a new set of technologies that some claim could shift the locus of control from institutions to individuals.

## 3. Both Facilitate Control over the Individual to Enable Extractive Practices on Behalf of Institutions.

Both the Blockchain Pilot Project and the Nansen Passport could ultimately facilitate control over individuals and enable extractive practices on behalf of institutions. Embedded in all forms of identification is the potential danger to its bearer. Threats to the security of refugees and migrants, however, deserve special consideration. Their vulnerability often occasions dangerous actions in order to remain invisible to institutions. Attempts by refugees to elude data collection have engendered such extreme measures as the burning of fingertips to prevent biometric registration by the EURODAC system (Gillespie et al. 2018). This phenomenon, termed by some 'systems avoidance,' evinces the seriousness of the threat posed by identification (Brayne 2014). The entanglement of the Blockchain Pilot Project with private sector partners including Safaricom M-Pesa raises concerns of data sharing and abuse of information access, potentially creating tremendous negative downstream effects of technology. Other agreements struck by humanitarian organizations, such as the data sharing agreement between UN World Food Programme (WFP) and Palantir, raise concerns, generally, about the collaboration between the public and private sectors on issues of digital identification. These collaborations could enable extractive data practices on behalf of corporations and governments, facilitate control over the individual, and, possibly most concerning of all, endanger individuals.

The Nansen Passport similarly enabled the extension of institutional control over individuals. The very existence of the Nansen Passport that had only been conferred on two groups—'personnes d'origines Russe' and Armenians—served to distinguish them as a danger to states, marking them for sustained consideration at the border. Even the Russians, who Arendt once wrote were 'in every respect the aristocracy of the stateless,' were regarded with suspicion as potential agents of Bolshevism (Arendt 1968). Indeed, in the wake of the Russian plight, the United States secured its borders, discouraging entry for individuals from the Russian Empire (Kerber 2007). Armenian refugees were regarded with similar suspicion (Tixier 1925). The Nansen Passport thus served to segregate persons without nationality from other travelers. This served to enhance state control over movement.

## 4. Both Began as Efforts to Render Communities 'on the Periphery' of Global Power 'Legible.'

Though both the Blockchain Pilot Project and the Nansen Passport began as attempts to extend identification to individuals 'on the periphery,' they also occurred coextensively, and may even prefigure (it remains too early to tell with digital identification) with attempts to extend identification at the center of the nation-state system.

Beyond attempting to empower vulnerable populations, the formation of the Nansen Passport was also entangled with the formalization of the Modern International Passport. Most accounts of the Modern International Passport contend that at the 1926 Organization for Communications and Transit Passport Conference, the Passport was standardized and globalized (Turack 1972; Torpey 2000; Salter 2003). I argue that the formation of the Nansen Passport was entangled in the formalization of the Modern International Passport.

The primary aim of the 1926 Conference was to abolish all passports (at the time, myriad national passport regimes coexisted), in order to reestablish pre-war conditions of relatively free movement (League of Nations 1926). Most characterize the pre-war period as an era in which general prosperity produced freedom of movement for most. This came to an abrupt end with the arrival of World War I (WWI) (Hobsbawm 1994). According to Polanyi (1944), the War was characterized by a 'crustacean-style nationalism,' nations crabbily distinguished between 'us' and 'them.' To conserve their populations for conscription and taxation and secure their borders, states intensified controls on movement for citizen and non-citizen alike. Though several delegates at the 1926 Conference attempted to repeal all passport systems, this aim was held in abeyance by the existence of three groups: persons without nationality, persons of doubtful nationality, and emigrants.

Each group represented a threat to states. Persons without nationality and persons of doubtful nationality were viewed with suspicion; several delegates expressed concern that refugees from Russia, for instance, would spread bolshevist ideology (League of Nations 1926). Emigrants, likewise, were regarded with trepidation; WWI had decimated many state economies, engendering an unwillingness to import foreign labor (Hobsbawm 1987). In this climate of protectionism, emigrants were seen as a threat to state economies. For these reasons, many states interpreted persons without nationality, persons of doubtful nationality, and emigrants as hazards.

Though delegates endeavored to abrogate all passports, the existence of these three groups necessitated a means of controlling the movement of some. The Conference considered several mechanisms to exclusively restrict the movement of these groups. Among them were the globalization of the Nansen Passport for persons without nationality and persons of doubtful nationality, the inscription of an 'E' on the frontispiece of passports for emigrants, and the creation of a distinct 'identity book' for emigrants. None, however, were ultimately accepted because each was seen as impractical. The provision of identification technology for some but not others was seen as a prohibitive logistical challenge (League of Nations 1926). Instead of only using identification technology for these groups, delegates made use of two identification technologies coextensively. In this way, the formation of the Nansen Passport and the formalization of the Modern International Passport were entangled.

Following the Conference, some states used the Nansen Passport as an alternative to the Modern International Passport for persons without nationality and persons of doubtful nationality. The Refugee Passport was first proposed as an alternative to the Modern International Passport by the German delegation in 1924 at the International Conference of Emigration and Immigration. Germany argued that the Refugee Passport should be extended universally to all persons without nationality (Turack 1972). Though the German delegation's proposal was rejected, by 1926 all represented governments, with the exception of Latvia, agreed to recognize the Refugee Passport as an alternative to the Modern International Passport for persons without nationality on a group-by-group basis.

At the time it had already been conferred on two groups: 'personnes d'origines Russe' and Armenians no longer under the protection of the Turkish Republic (White 2017). It served to distinguish them to states, marking them for sustained consideration at the border. In 1927, the Nansen Passport was extended further to 'persons of doubtful nationality.' Though a contentious phrase, delegates ultimately defined 'persons of doubtful nationality' as individuals on whom consular authorities refused to confer national passports (League of Nations 1926). Though they were not stateless, nor without nationality, the Refugee Passport was recognized by some states as an alternative to the Modern International Passport for persons of doubtful nationality. For persons without nationality and persons of doubtful nationality, therefore, some states recognized the Nansen Passport as an alternative to the Modern International Passport.

The Refugee Passport was also extended, by some states, to emigrants as an alternative to the Modern International Passport. Emigrants were a controversial category at the 1924 and 1926 Conferences (Turack 1972). Indeed, even the question 'what is an emigrant' invited debate. The matter was not settled until the delegate from the United States defined an emigrant as, in effect, any individual seeking durable residence in a nation not their own (League of Nations 1926).

At the International Conference of Emigration and Immigration hosted in Rome in May 1924 delegates attempted to repeal the use of passports for all travelers *but* emigrants (Turack 1972). The Italian delegation summarized the reason for their exclusion succinctly: 'whereas before [the War] the wave of emigrants could go anywhere now entry depends on the economic conditions in-country' (cited in Turack 1972). WWI had decimated national economies, inciting panic throughout Europe (Hobsbawm 1987). Even at the 1926 Conference, eight years after the armistice, states continued to argue they could not sustain larger workforces (League of Nations 1926). In this period of protectionism, some states perceived emigrants as threats to national economies. Much like persons without nationality and persons of doubtful nationality, emigrants had to be marked for further consideration and possible exclusion.

The Refugee Passport was first proposed for emigrants by the representative of Estonia: 'Passports for emigrants [should be] granted in conformity with the decisions taken at the conference called by Dr. Nansen at Geneva, July 3–July 5, 1922,' the conference at which the Refugee Passport entered into effect (League of Nations 1926). It was the second time the Refugee Passport had been considered in relation to citizens (the first being for the aforementioned persons of doubtful nationality). Some states extended the Refugee Passport, thereafter, to persons of doubtful nationality and emigrants (League of Nations 1926). The Refugee Passport served to distinguish the three groups from other travelers. Carriers of the Refugee Passport were effectively marked for restriction. By enhancing restrictions on some, states were able to facilitate movement for others. Limitations on movement for persons without nationality, persons of doubtful nationality, and emigrants made possible relatively unrestricted movement for carriers of the Modern International Passport, thus achieving many of the delegates' aim of reestablishing conditions of (relatively) free movement for some. In this way, the formation of the Nansen Passport was entangled with the formalization of the Modern International Passport. Thus, not only was the Nansen Passport extended to vulnerable individuals in an attempt to empower them, its formation was also entangled in the formalization of the Modern International Passport. Though it is too early to tell what the results of the extension of digital identification to vulnerable populations will be, it is possible that it will similarly affect non-vulnerable populations.

## CONCLUSION

Though fixtures of contemporary debate in technologically 'advanced' societies, topics of data, security, and identification are rarely considered in relation to vulnerable populations (Lyon 2007). It has been one aim of this chapter to generate a discussion on the extension of digital identification to vulnerable populations. In this spirit, I examined two identification technologies separated by roughly a century in order to couch an examination of digital identity in relation to a historical identification technology. In the process, I drew out several commonalities between the two identification technologies: both received justification on the basis of extending rights and services, were produced through a collaboration between the public and private sectors, made possible control over individuals, and worked to render communities 'on the periphery' of global power 'legible.'

The globalization of digital identification could have dramatic implications for vulnerable and non-vulnerable populations alike. Today, roughly 1.1 billion individuals lack an officially recognized form of identification (World Bank 2018). Without a recognized form of identification, these individuals are unable to access government services and humanitarian aid. Obtaining identification can be time-intensive and cost-heavy. This gap is recognized by the United Nations Sustainable Development Goal 16.9, which seeks to provide legal identity for all by 2030. Several organizations, governmental and non-governmental, are working towards this ambitious goal.

As a lever of governance, digital identification could streamline the distribution of services and analysis of data. Currently, many national services are fragmented. Departments of governments operate independently. This can lead to system inefficiencies and overlap. The same problem is common to the Humanitarian Sector, where interagency coordination is rare, especially in non-disaster conditions. Constructing a single, unified digital identification could facilitate the extension of services for both governmental and non-governmental organizations. However, digital identification could also be used as a tool of control, surveillance, and extractive data practices.

The future of identification technologies for both vulnerable and non-vulnerable populations remains unclear. What is certain, however, is that their fates are entangled. As I have demonstrated, two pairs of identification technologies separated by roughly a century both evince intricate ties. Those considering questions of data, identification, and security would do well to recognize this fact and take note of the work already being done in the Humanitarian Sector to provide answers.

## BIBLIOGRAPHY

Accenture. 2017. Accenture, Microsoft Create Blockchain Solution to Support ID2020 (accessed May 2, 2010).

Accenture News Release. Available from: https://newsroom.accenture.com/news/accenture-microsoft-create-blockchain-solution-to-support-id2020.htm (accessed April 4, 2010).

Agamben, G. 1998. *Homo Sacer: Sovereign Power and Bare Life* (Meridian: Crossing Aesthetics Series). Stanford, CA: Stanford University Press.

Arendt, H.1968. *The Origins of Totalitarianism.* New York and London: Harcourt Brace and World.
Baraniuk. B. 2018. Bitcoin Electricity in Iceland Set to Overtake Homes. BBC. Available from: https://www.bbc.co.uk/news/technology-43030677 (accessed June 1, 2010).
BBC. 2017. UK Data Protection Laws to be Overhauled. Available from: https://www.bbc.co.uk/news/technology-40826062 (accessed June 1, 2010).
Bedoya, A. 2014. Big Data and the Underground Railroad. Slate. Available from: http://www.slate.com/articles/technology/future_tense/2014/11/big_data_underground_railroad_history_says_unfettered_collection_of_data.html Google Scholar (7th November).
Brayne, S. 2014. Surveillance and System Avoidance: Criminal Justice Contact and Institutional Attachment. *American Sociological Review*, 79(3): 367–91.
Brubaker, R. 1992. *Citizenship and Nationhood in France and Germany.* Cambridge, MA: Harvard University Press.
Carter, D. 1994. The Art of the State: Difference and Other Abstractions. *Journal of Historical Sociology*, 7(1): 73–102.
Dillon, M. 1998. The Scandal of the Refugee: Some Reflections on the 'Inter' of International Relations Refuge. *Canada's Journal on Refugee*, 17(6): 30–39.
Dinkins, D. 2017. Collapse of Bitcoin's 'New York Agreement' Would Have Long-term Consequences. *Coin Telegraph.* Available from: https://cointelegraph.com/news/opinion-collapse-of-bitcoins-new-york-agreement-would-have-long-term-consequences
Dybaeker, R. 2001. Verification Versus Validation: A Terminological Comparison. *Accreditation and Quality Assurance*, 16(2): 105–8.
Economist, The. 2015. The Promise of the Blockchain; the Trust Machine. Available from: https://www.economist.com/leaders/2015/10/31/the-trust-machine (accessed May 5, 2010).
Ellerman, A. 2010. Undocumented Migrants and Resistance in the Liberal State. *Politics & Society*, 38(3), 408–29.
GDPR. 2018. General Data Protection Regulation. Available from: https://gdpr-info.eu/ (accessed June 5, 2010).
Gillespie, M., Osseiran, S. and Cheesman, M. 2018. Syrian Refugees and the Digital Passage to Europe: Smartphone Infrastructures and Affordances. *Social Media and Society* (20 March). doi: https://doi.org/10.1177/2056305118764440.
Gordenker, L. 1987. *Refugees in International Politics*. London: Croom Helm.
Gordon, D. The Electronic Panopticon: A Case Study of the Development of National Criminal Records System. *Politics & Society*, 15(4): 483–511.
GSMA. 2017. Blockchain for Development. GMSA Reports. Available from: https://www.gsma.com/mobilefordevelopment/wp-content/uploads/2017/12/Blockchain-for-Development.pdf (accessed May 6, 2010).
Habermas, J. 1998. *The Inclusion of the Other*. Cambridge: Polity Press.
Haddad, E. 2008. *Refugee in International Society: Between Sovereigns.* Cambridge: Cambridge University Press.
Haggerty, K.D. and Ericson, R.V. 2003. The Surveillant Assemblage. *The British Journal of Sociology*, 51(4): 605–22.

Hathaway, J.C. 2011. *Law of Refugee Status*. Melbourne: Cambridge University Press.
Hieronymi, O. 2003. The Nansen Passport: A Tool of Freedom of Movement and of Protection. *Refugee Survey Quarterly*, 22(1): 36–7.
Hobsbawm, E. 1987. *The Age of Empire, 1875–2012*. London: Abacus.
Hruska, J. 2014. One Bitcoin Group Now Control 51% of the Total Mining Power. Available from: https://www.extremetech.com/extreme/184427-one-bitcoin-group-now-controls-51-of-total-mining-power-threatening-entire-currencys-safety
Hyndman, J. 2000. *Managing Displacement: Refugees and the Politics of Humanitarianism*. Minnesota, MI: University of Minnesota Press.
ID2020. 2017. ID2020. Available from: https://id2020.org/news/http/id2020org/news/2017/9/28/id2020-update-from-the-2017-united-nations-general-assembly (accessed May 6, 2010).
ID2020. 2018. ID2020. Available from: https://id2020.org/partnership/ (accessed May 6. 2010).
Juskalian, R. 2018. Inside the Jordan Refugee Camp That Runs on Blockchain. MIT Technology Refugee. Available from: https://www.technologyreview.com/s/610806/inside-the-jordan-refugee-camp-that-runs-on-blockchain/ (accessed June 5, 2010).
Kakuma Daily News Reflector. 2016. Refugee Verification Excercise. Available from: https://kanere.org/2016/12/31/refugee-verification-exercise/ (accessed May 5, 2010).
Kerber, L.K. 2007. The Stateless as the Citizen's Other: A View from the United States. *American Historical Review*, 112(1), 1 February, 1–34.
Kristeva, J. 1988. *L'Etrangers a Nous-Memes*. Paris: Librairie Artheme Fayard.
Kristeva, J. 1991. *Strangers to Ourselves*. New York: Columbia University Press.
Kulischer, E. 1943. *The Displacement of Population in Europe*. Montreal: International Labour Office.
Latonero, M. and Kift, P. 2018. On Digital Passages and Borders: Refugees and the New Infrastructure for Movement and Control. *Social Media and Society*, 4(1).
League of Nations Archive. 1922. General Report on the Work Accomplished up to March 15 1922 by Dr Fridtjof Nansen, High Commissioner of the League of Nations. U.N.F/11 C.124.M.74.1922, 15 March 1922.
League of Nations Archive. 1926. Organization for Communications and Transit Passport Conference, the Modern International Passport July 6 1926 by Organization for Communications and Transity of the League of Nations. U.N.F/11 C. 423. M. 156. 1926 VIII.
Lloyd, M. 2003. *Passport: The History of Man's Most Travelled Document*. Gloucestershire: Sutton Publishing.
Long, K. 2013. When Refugees Stopped Being Migrants: Movement, Labour, and Humanitarian Protection. *Migration Studies*, 1(1): 4–26.
Lyon, D. 2007. Surveillance, Power, and Everyday Life. Available from: https://kanere.org/2016/12/31/refugee-verification-exercise/
Malkki, L. 1995a. *Purity and Exile: Violence, Memory, and National Cosmology among Hutu Refugees in Tanzania*. Chicago, IL: University of Chicago Press.

Malkki, L. 1995b. Refugees and Exile: From 'Refugee Studies' to the National Order of Things. *Annual Review of Anthropology*, 24(1): 496–523.
Martin, L.H., Gutman, H. and Hutton, P.H. (eds). 1988. *Technologies of the Self: A Seminar with Michel Foucault*. London: Tavistock.
Marx, G.T. 1998. Ethics for the New Surveillance. *The Information Society*, 14: 171–85.
MasterCard. 2018. Digital Identity. Available from: https://www.mastercard.us/en-us/governments/find-solutions/financial- inclusion/digital-identity.html (accessed May 5, 2010).
Marrus, R. 1985. *The Unwanted: European Refugees in the Twentieth Century*. New York: Oxford University Press.
McKeown, A. 2008. *Melancholy Order: Asian Migration and the Globalization of Borders*. New York: Columbia University Press.
Morley, F. 1932. *The Society of Nations: Its Organization and Constitutional Development*. Washington, DC: Brookings Institution.
Morozov, E. 2014. *To Save Everything, Click Here: The Folly Solutionism*. London: PublicAffairs.
Noiriel, G. 1996. *The French Melting Pot: Immigration, Citizenship, and National Identity*. Minnesota, MI: University of Minnesota Press.
Polanyi, K. 1944. *The Great Transformation*. New York: Farrar & Rinehart.
Poster, M. 1990. *The Mode of Information: Poststructuralism and Social Context*. Chicago, IL: University of Chicago Press.
Rahman, Z. 2017. Irresponsible Data? The Risks of Registering the Rohingya. Available from: https://www.irinnews.org/opinion/2017/10/23/irresponsible-data-risks-registering- rohingya (accessed May 4, 2010).
Ramonat, S. From Identity to Refugee Status Determination. ID2020DesignWorkshop github. Available from: https://github.com/WebOfTrustInfo/ID2020DesignWorkshop/blob/master/topics-and-advance-readings/identity_to_refugee_status_determination.md
Robertson, C. 2015. *The Passport in America: The History of a Document*. New York: Oxford University Press.
Rose, N. 1996. *Powers of Freedom*. Cambridge: Cambridge University Press.
Salter, M. 2003. *Rights of Passage: The Passport in International Relations*. Boulder, CO: Lynne Rienner.
Sargasan, M. 2017. India's ID System Is Reshaping Ties between State and Citizens. *The Economist*. Available from: https://www.economist.com/asia/2017/04/12/indias-id-system-is-reshaping-ties-between-state-and-citizens
Satariano, A. 2018. New Privacy Rules Could Make This Woman One of Tech's Most Important Regulators. *New York Times*. Available from: https://www.nytimes.com/2018/05/16/technology/gdpr-helen-dixon.html (accessed June 5, 2020).
Scott, J. 1998. *Seeing Like a State: How Certain Schemes to Improve the Human Condition Have Failed*. New Haven, CT: Yale University Press.
Selby, J. 2004. *Water, Power and Politics in the Middle East: The Other Israel-Palestine Conflict*. London: I.B.Tauris & Company.
Skran, C. 1995. *Refugees in Inter-war Europe: The Emergence of a Regime*. Oxford: Oxford University Press.

Soguk, N. 1999. *States and Strangers: Refugees and Displacements of Statecraft*. Minnesota, MI: University of Minnesota Press.

Sovrin. 2018. A Protocol and Token for Self-sovereign Identity. Available from: https://sovrin.org/library/sovrin-protocol-and-token-white-paper/

Tixier. 1925. *The Refugee Problem in Bulgaria*. Geneva: International Labour Organisation, p. 25.

Torpey, J. 2000. *The Invention of the Passport: Surveillance, Citizenship and the State*. New York: Cambridge University Press.

Turack, D. 1972. *The Passport in International Law*. Lexington, MA: Lexington Books.

UNHCR. 2015. Joint Inspection of the Biometrics Identification System and Food Distribution in Kenya. Available from: https://documents.wfp.org/stellent/groups/public/documents/reports/wfp277842.pdf (accessed May 5, 2010).

UNHCR. 2017a. The UNHCR Nansen Refugee Award Celebrates Heroes Supporting Refugees, Internally Displaced and Stateless People. Available from: http://www.unhcr.org/uk/nansen-refugee-award.html (accessed May 5, 2010).

UNCHR. 2017b. ID2020 and UNHCR Host Joint Workshop on Digital Identity. UNHCR Blogs. Available from: http://www.unhcr.org/blogs/id2020-and-unhcr-host-joint-workshop-on-digital-identity/ (accessed May 5, 2010).

UNHCR. 2018. Bridging the Identity Divide—Is Portable User-centric Identity Management the Answer? UNHCR Blogs. Available from: http://www.unhcr.org/blogs/bridging-identity-divide-portable-user-centric-identity - management-answer/

Vota, W. 2018. A Really Bad Blockchain Idea: Digital Identity Cards for Rohingya Refugees. NYU Tandon School of Engineering. Available from: http://thegovlab.org/a-really-bad-blockchain-idea-digital-identity-cards-for-rohingya- refugees

Wang, V. and Tucker, J. 2017. Surveillance and Identity: Conceptual Framework and Formal Models. *Journal of Cybersecurity*, 3(3): 145.

White, E. 2017. The Legal Status of Russian Refugees, 1921–1936. Comparativ. Zeitschrift fur Globalgeschichte und Vergleichende Gesellshaftsforschung. University of West England: 18–38.

Wittgenstein, L. 2007. *Tractatus Logico-Philosophicus*. New York: Cosimo.

World Bank Identification for Development. 2018. ID4D Global Dataset. Available from: https://id4d.worldbank.org/global-dataset (accessed January 2021).

World Food Programme. 2018. Building Blocks. World Food Programme Innovations. Available from: http://innovation.wfp.org/project/building-blocks (accessed May 5, 2010).

Zetter, R. 1991. Labelling Refugees: Forming and Transforming a Bureaucratic Identity. *Journal of Refugee Studies*, 4(1): 39–62.

Zetter, R. 2007. More Labels, Fewer Refugees: Remaking the Refugee Label in an Era of Globalization. *Journal of Refugee Studies*, 20(2): 172–92.

Zolberg, A. 1983. The Formation of New States as a Refugee-generating Process. *The Annals of the American Academy of Political and Social Science*, 467: 24–38. Available from: http://www.jstor.org/stable/1044926

Zolberg, A. 1999. Matters of State: Theorizing Immigration Policy. In C. Hirschman, P. Kasinitz and J. De Wind (eds), *The Handbook of International Migration: The American Experience*. New York: Russell Sage, pp. 71–93.

# 4. Politics of technology: the use of artificial intelligence by US and Canadian immigration agencies and their impacts on human rights

**Roxana Akhmetova and Erin Harris**

## INTRODUCTION

Over the past couple of decades, there has been a global rise in the use of new and emerging border technologies. Following a trend by a majority of countries in the Global North to securitize migration, the ways in which certain types of migration, such as asylum, are understood and managed are vastly changing. The US and Canada are at the forefront of incorporating artificial intelligence (AI) and machine learning (ML) technologies into border control and migration governance; however, the extent to which the two governments are incorporating and testing new AI and ML technologies varies. Regardless, both countries are showing an ever-increasing level of interest to incorporate AI and ML technologies into their surveillance, border control, and decision-making.

The US has been incorporating numerous smart border technologies such as aerial and satellite surveillance, algorithmic risk assessment through AI and ML processing, as well as fiber-optic sensor systems. While each border technology is used for various purposes, the conglomeration of new and emerging technologies which heighten the presence of surveillance, in addition to the physical border wall, has led to massive changes on how asylum cases are handled and processed. Furthermore, border technologies are only increasing and becoming ever more lucrative by giving more authority to government agencies, collaborating with bordering countries to expand policing beyond the border, and stripping migrants of many of their basic human rights before even arriving at the border. The use of these technologies goes beyond border control and also includes decision-making on asylum applications. AI and ML promise to revolutionize how government agencies do their work. Recent developments in AI and ML have the potential to not only secure but make greater use of administrative data, improve the quality of decisions, and reduce the cost of core governance functions, thus making governments more efficient, accountable, and effective. Since many of the AI and ML-powered technologies are in their nascence, their use in governance presents a significant impact on the human rights of vulnerable groups of people like asylum seekers, stateless people, and minorities.

The main argument of our chapter is that the use of AI immigration enforcement technologies infringes upon humanitarian migrants' rights of privacy and protection, while increasing discrimination to diminish agency. We define humanitarian migrants as migrants who are seeking refuge or protection from their country of origin into either the US or Canada. We begin the chapter by discussing how the US and Canada have been incorporating AI into border control, surveillance, and government decision-making. In the US, AI has been solely used in surveillance technologies to analyse risks assessment of migrants but could expand into using AI in a more automated processing practice. Canada has been focusing on increasing the implementation of AI technologies into its immigration decision-making for at least the past five years. While the use of these technologies offers some benefits, the lack of a robust legal framework poses a number of concerns for the protection of human rights of vulnerable groups of people like asylum seekers, humanitarian migrants, and marginalized groups of people. We conclude with a series of recommendations for the US and Canadian governments as they continue to expand their use of AI not only in border control, surveillance, and immigration spheres.

## BIOPOLITICS, INVISIBLE BORDER WALLS, AND PRIVATIZATION OF IMMIGRATION

AI is the programming and training of a computer using statistical models to do tasks typically reserved for human intelligence (Calo, 2017). Computer algorithms are powerful tools for automating many aspects of life, especially those that require step-by-step routines, such as organizing and digitizing operational and administrative tasks, making them more consistent and faster (Bansak et al., 2018). ML is an application of AI that uses data to automatically learn and improve from experience without being explicitly programmed to perform some tasks by a human programmer (Flach, 2012). Recently available data and computing power have reached a point where it has become very useful to develop ML which can pick up patterns that humans may otherwise miss (UKGOS, 2016). AI and ML algorithms draw on vast amounts of data to learn and make inferences about patterns and future behavior, which has great potential in forecasting, managing, and controlling migratory flows but also mass surveillance (Beduschi, 2018; Rango, 2015).

As countries like the US and Canada have been pushing for the securitization of migration in their rhetorical discourse over the past couple of decades, the rise in the use of new and emerging technologies is transforming the way in which contemporary asylum cases are understood and managed. Biopolitics is defined as a way in which political power is exercised by the state through every aspect of human life as a way to maintain order on whole populations (Foucault, 1977–78). Moreover, biopolitics in the context of border security practices is further understood as a means to securitize migration. Expanding upon his theory, Foucault (2007) argues a new biopolitics in which technology is inherently political, by which he claims, "technologies of power have become more complicated, more meticulous, more subtle and neither less violent or more humane" (Jacobsen, 2016: 41). In applying Foucault's theory of a new biopolitics to border security practices (Jacobsen, 2016: 41), AI surveillance and ML are implicitly designed as subtle yet inhumane power-seeking tools. By using Foucault's theory of new biopolitics, we can begin to understand how the use of border technologies is intended to provide objectivity into the decision processing of asylum applications. However, AI immigration enforcement technologies are inherently political and often used to exploit migrants at various levels of segregation and vigilant policing. The biopolitics of border security technology is seen in how states use these technologies to control the rhetorical narrative of asylum and how an individual internalizes this in the state's attempt to control migrants in a way that serves the government. Thus, migrants who are seeking asylum are at the mercy of new government-sanctioned border security technologies that can easily be mismanaged as they relieve states of their responsibility to be accountable for any such mishandlings of applications.

Despite the fact that Foucault's theory was written nearly 40 years ago, it is still very much applicable to how governments use AI-powered technologies. In attempts to create "smart borders," we argue that both Canada and the US are in fact using AI technologies to erect "invisible border walls" that are aimed at diminishing government responsibility and accountability for their actions and decisions as well as to hide a divisive agenda to refuse entrance to undesirable migrants. Perhaps another motivation for the increasing use of new border technologies are the multi-layered interests of private technology and security and arms companies. Palantir Technologies is a private US software company, founded in 2003 by Peter Thiel, that specializes in big data analytics. The company is known for its notorious collaborations on projects to mine massive bits of data for security purposes with a number of US federal agencies such as the National Security Agency (NSA), the Federal Bureau of Investigation (FBI), and the Central Intelligence Agency (CIA). Over more recent years, the additional collaboration with US government agencies such as the US Customs and Border Protection (CBP) and the US Immigration and Customs Enforcement (ICE) has raised further controversies. Not only is Palantir acquiring a massive profit from the government, but such technologies are being used in classified operations that allow for many of their products to go unchecked through auditing and ethics committees (Greene, 2019b). While Palantir is a major contributor in supplying the government with border enforcement technologies, DHS incorporates a range of technologies from other security defense contractors, all potentially leading towards a corporatized border.

As reported in a recent investigation by Greene, the US government currently has over 29 active contracts with Palantir alone that are worth approximately $US1.5 billion (Greene, 2019a). Palantir is responsible for building case management software for the US Department of Homeland Security (DHS) agencies such as ICE and CBP (Mijente, 2018: 3). Palantir's AI and ML software take large data sets collected from surveillance technologies and aggregate the data accumulated to identify any suspicious trends and deter any migrants that, as a result, are flagged as suspicious threats upon their arrival at the border. Greene reports that after Google abandoned Project Maven, a controversial program initiated by the Pentagon to "build an AI-powered surveillance platform for unmanned aerial vehicles," Palantir agreed to continue with the project despite concerns over ethics (Greene, 2019b). While Greene (2019b) reports the figure the firm is receiving from Project Maven is approximately $US10 billion, Peterson (2020) contends that it is as "important as America's race to develop a nuclear weapon during World War II," and thus Palantir is likely to receive a similar, if not a greater, amount of money.

Palantir has also been trying to branch out into Canada for some time, although its growth has been slow. Palantir partnered up with the

Canadian Department of National Defense to provide data analytics software for the Canadian Forces Special Operations Command and the Calgary Police Department to integrate their database (Braga, 2017; Hemmadi, 2019). Palantir's entrance into the Canadian market is perfectly timed, as Ottawa is trying to increase its cyber capabilities in partnership with its NATO and "Five Eyes" allies (Canada, the US, the UK, Australia, and New Zealand). Palantir does not mind being closely tied to governments despite a number of high profile tech firms, like Google and Microsoft, marking their independence from governments (MacMillan and Dwoskin, 2019). Shortly after ending his post as the Canadian ambassador to the United States (2016–19), David MacNaughton became the head of the Canadian wing of Palantir. Throughout his tenure as the ambassador to the United States, MacNaughton was the trusted advisor to the Canadian Prime Minister Justin Trudeau. The link between Palantir and the Prime Minister's Office was further cemented with the re-election of Justin Trudeau in 2019, which may prove to be very useful for the company. A partnership with Trudeau, who represents a brighter public friend than Donald Trump and ICE, might be exactly what Palantir needs to boost its popularity and re-brand itself.

Although it is not clear what work is being done between Canada and private companies, Palantir would be a natural fit for Canada's newly established Communications Security Establishment (CSE), which recently gained the authority to conduct more active cyber operations at home and abroad in 2019. As such, a partnership with Palantir might be a good fit for the technologies CSE is using to track individuals and interpret data. Palantir is not the only private company to have an office in Canada, and it most likely will not be the last. A UAE-based private tech company that focuses on cybersecurity, DarkMatter, also set up an office in Toronto, Canada, in 2016 (DarkMatter, 2016). DarkMatter is also a controversial company and is currently under investigation by the FBI for engaging in crimes related to espionage, murder, and incarceration of foreign nationals (Mazzetti et al., 2019). Fears around having these and other companies, like Huawei, raise concerns not only about who is responsible for data protection, but also which governments these companies are loyal to.

Despite major concerns over the vast technologies acquired by the DHS in securitizing migration, there has been very little dialogue addressing the erosion of checks and balances as a result of the lucrative collaborations between private firms like Palantir and the US government. This perhaps may be due to the high level of classification of operations like

Project Maven, where very little information on the AI surveillance program is open to the public. Alex Karp, the CEO of Palantir, contends that governments are responsible for the protection of data as well as the assurance of privacy and lawful access of data. In a recent interview, Karp argued that "the present and future ability to control the law and its application will be determined by our ability to harness and master artificial intelligence and its predecessor, machine learning" (Bloomberg Politics, 2019). Although a slightly more progressive figure than Palantir's founder, Peter Thiel, Karp remains firm in his belief in the importance between private firms and their collaboration with government security agencies. Karp and the use of AI and ML-backed surveillance are two of many novel security and defense technologies acquired in what is often dubbed the "smart wall" – a virtual expansion of border security. Despite bipartisan support over a "smart wall," a main concern is that incorporating new smart border technologies will "enrich tech and military contractors while violating our civil rights" (Franco, 2019). Due to most of the AI and ML-powered operations being classified, the use of AI and ML software infrastructure implemented by the DHS in testing pilot programs has severe ethical implications on how asylum law is administered and adjudicated. In a recent DHS audit report, the US Inspector General found that CBP lacked numerous measures of safeguards for information and data collection (US Department of Homeland Security, 2018). More importantly, it has become relatively unclear who bears responsibility when a program leads to faulty or erroneous processing of asylum applications due to a lack of proper safeguards in piloted infrastructure.

Several Canadian cities like Toronto and Calgary have also been using facial recognition technology, since at least 2018, while Montreal and Halifax have neither admitted nor denied the use of AI-powered surveillance technologies (Lee-Shanok, 2019). Canadian privacy laws and judicial authorization required by the Criminal Code have been preventing Canadian government agencies from gathering big data and conducting expansive surveillance of its citizens to the same extent as the US. The issue of increasing the role of private companies, like Palantir, is that they can change this trend. Outsourcing policing, surveillance, and decision-making to private companies raises issues around the role of companies in governance, accountability, and transparency. Despite Canada's status as an AI and ML innovation hub, its government has yet to develop a regulatory regime to deal with issues of discrimination, accountability, and the role of domestic and foreign technology and cybersecurity companies. A lack of a legal framework that specifically targets how AI and ML are used, how the data is collected and treated, who has access to it, and who is accountable for the issues that arise from these technologies undermines trust in these technologies. This is especially troubling as the use of facial recognition is being used by Canadian law enforcement agencies in public spaces. Oversight over the use of AI and ML should be divided among federal departments and agencies in order to make regulations specific to each industry.

## CANADA'S USE OF AI AND ML

Since at least 2014, the Canadian federal government has been testing automated decision-making systems powered by AI and ML to automate activities that are typically performed by immigration officials to support the evaluation of some migrant and visitor applications. The transition from test to use of AI and ML are on the way, including decisions on whether an applicant should be designated as a risk to Canada (Molnar and Gill, 2018). Some of the uses of AI and ML are encouraging; however, much of this technology is in its nascence and is very experimental in nature. ML is used through AI by automatically learning data and improving from experience without being explicitly programmed to perform some tasks by a human programmer (Flach, 2012). The use of these technologies has alarming implications for the fundamental human rights of those subjected to their use and there is a need to consider the implications of increased reliance of AI and ML by Canadian government agencies. A lack of a legal framework to guide Canadian government agencies' use of AI and ML can have two interrelated implications: accountability issues and abuse of human rights. Canada's use of AI is not as advanced and expansive as that of the US; however, a number of AI technologies are being tested and already implemented in the Canadian immigration system.

Canada's immigration system is federally regulated by the Ministry of Immigration, Refugees, and Citizenship Canada (IRCC) and falls under administrative law. It is governed by the Immigration and Refugee Protection Regulations (IRPR), the Immigration and Refugee Protection Act (IRPA), and internal operational documents of various affiliate branches. Every initial immigration decision is made either by an administrative tribunal like the Immigration and Refugee Board (IRB), by individual immigration officers that work for IRCC, or by the Canadian Border Services Agency (CBSA) which is an enforcement arm of immigration. Canada has two types of international protection programs: the

Refugee and Humanitarian Resettlement Programs (RHRP) and the In-Canada Asylum Program (ICAP). The ICAP is for those who begin their asylum procedure in Canada at the border or when they are already in the country (Government of Canada, 2019). The RHRP is for individuals who were granted the status of a refugee and are hosted in Canada (Government of Canada, 2019). The RHRP has three sub-categories depending on who is sponsoring the refugee: government, private, and blended sponsorship. These sponsorship programs provide refugees with support for one year, or until the refugee becomes economically independent. At the end of the sponsorship, the refugee is eligible for aid provided by their province/territory or municipal government.

Asylum claimants whose application was rejected by the Refugee Protection Division (RPD) may appeal to the Refugee Appeal Division (RAD) in order to assess if the decision was wrong based on law, the facts, or both (Government of Canada, 2019). These decisions are reviewed by an appeals body like the RAD, the Federal Court of Canada, or the Federal Court of Appeal, before moving up to the Supreme Court of Canada. At every stage of an individual's immigration proceeding, many decisions are made about their application. RAD decides whether to confirm or change RPD's decision or send the case back to the RPD for another hearing. RAD makes its decisions without a hearing, based on submissions and evidence, which may be new, that the parties provide. RAD may order an oral hearing to consider the new evidence submitted. Since using AI by the Canadian federal government is a relatively novel endeavor, it is not yet clear how the right to appeal a decision will be upheld during the use of automated decision-making systems. Using AI and ML in the immigration sphere presents ethical and technical problems. First, asylum and immigration-related evaluations are typically qualitative in nature and are performed by humans. Using AI and ML in immigration areas can be particularly ethically challenging not only because these applications are the most discretionary in Canada's immigration system, but also because their decisions have significant impacts on the lives of individuals seeking refuge in Canada. Considering that the subjects of humanitarian and refugee applications are some of the most vulnerable individuals in the Canadian immigration system, they should be the last cases to be used in pilot programs.

Despite the international community praising Canada's immigration program, discriminatory decision-making and screening practices continue to plague Canada's immigration system and border control. The use of AI and ML technologies in the immigration sphere promises to significantly reduce or even eliminate conscious and subconscious forms of human bias which lead to discretionary outcomes. If AI and ML can deliver on this promise, this would significantly improve government decision-making and help ensure that government decisions treat newcomers to Canada ethically and fairly. However, many of the AI and ML technologies that Canada uses are currently in their experimental stages, although some have passed this phase and are being used in real life. Despite this, their impacts, benefits, and drawbacks are not yet fully known. Bias and data-related issues are of great concern to how Canada uses AI and ML in the immigration space, in part because once a particular bias is part of the technology's algorithm, the machine will not only perpetuate the biased decision-making, but it can compound already entrenched disadvantages and even develop new ways of discrimination. The semi-autonomous nature of algorithms, the ability of AI and ML to diverge from their intended purpose poses significant challenges, especially if this problem goes undetected.

A number of issues surround the data that is fed into AI and ML systems as it may be colored by implicit and explicit bias. Algorithmic bias is when AI and ML technologies arrive at decisions that are discriminatory despite these technologies being designed to be impartial. Algorithmic bias can come not only from the technology designer's prejudices or the biases of the society-at-large, but issues can also arise from underlying trends within the data or if the data was collected in biased ways. For example, in 2017, the Royal Canadian Mounted Police (RCMP) faced heavy criticism for engaging in religious and ethnic profiling of migrants near the unofficial border crossing between the US and Quebec at Roxham Road (Peritz and Leblanc, 2017). Without any clear rationale, and supposedly on its own initiative, the RCMP collected questionnaires from over 5,000 asylum seekers, featuring questions clearly colored by Islamophobic stereotypes (Shepherd, 2017). The questionnaire sought information about social values, political beliefs, and religion – including questions related to the individual's perception of women who do not wear a hijab, their opinions on ISIS and the Taliban, as well as the number of times a day the individual prayed (Shepherd, 2017). The questions clearly targeted Muslim individuals crossing the border, as no questions were included about other religious practices or terrorist groups (Shepherd, 2017). The collected answers were entered into an RCMP database, which could be shared with CBSA and "other security partners" (Shepherd, 2017). The RCMP claims that the questionnaire ceased to be used following an investigation by the *Toronto Star*, a Canadian daily newspaper, in October 2017 and the 5,438 files have been redacted (Shepherd, 2017).

Further, a number of methodological difficulties haunt data collection for AI algorithms such as sampling bias. A lot of data is collected from internet use or mobile phone use, however, those who have a mobile phone and use social media platforms are not necessarily representative of the population at large; this is especially true for asylum seekers. Although an increasing number of migrants have cell phones, income/gender/age inequalities exist among migrants and thus influence the data collected and the type of inferences that can be based using this data. "Information precarity" has been used to describe the challenges facing migrants who may have inconsistent and costly access to technologies, lack control over their own data, and experience anxiety about phones being used for surveillance of their activities (Wall, 2020). This challenge also presents issues of making sense of copious amounts of data and the need for advanced analytical capacities to process and filter data. Where such capacities exist, private companies typically own them, and this also presents issues around accountability, privacy, and data ownership. Yet another example of unethical collection of data occurred in 2018, when it was reported that the CBSA used private third-party DNA services such as Ancestry.com to establish the nationality of individuals subject to potential deportation (Khandaker, 2018). This is deeply concerning, not only because of the coercive nature of privacy invasion, but also because one's DNA is not related to nationality and should bear no impact on one's application (Molnar and Gill, 2018). Even if the government is not currently using this method of data collection, DNA samples could provide us with new information in the future and it could be used by governments in ways that we cannot imagine today. An issue to be mindful of is the extent to which Canadian government agencies outsource technologies for surveillance and border control and buy technologies from private domestic and foreign companies.

Depending on the way the algorithm is designed to sort data and make decisions, it may result in unintended discrimination or negative feedback loops that reinforce and exacerbate existing inequalities (Ng, 2017). There are several sources of algorithmic bias, such as human bias, issues with the way data was collected, and bias autonomously generated by the algorithm. Once a bias is entered into the system, it can not only be compounded to create discriminatory outcomes that reproduce and magnify already existing biased practices, but algorithms can even develop their own way of discrimination. Due to the semi-autonomous nature of algorithms, they can diverge from their intended purpose; this issue can pose significant challenges especially if this problem goes undetected. The obscure nature of immigration and refugee decision-making creates an environment that can be perfect for algorithmic discrimination. Relying on AI to decide whether an asylum claimant's story is truthful to whether a potential immigrant's marriage is genuine can be highly discretionary and depend on the individual officer's assessment of credibility (Satzewich, 2014). An IRCC spokesperson stated that automated decision-making systems are built by analysing thousands of past applications and their outcomes. This allows the technology to "learn" by detecting patterns in the data in a way that is close to how officers learn through experience of processing applications (Keung, 2017).

The use of such technologies has significant implications for the fundamental rights of those subjected to them. Thus, proper use of AI by government agencies is gravely important. If managed well, AI and ML tools can modernize public administration and bureaucracy, resulting in more accurate, efficient, and equitable forms of state action. If these technologies are not managed properly, their use can result in undesirable decision-making, widen the public-private technology gap, increase the potential of arbitrary government action and power, enable surveillance that could threaten privacy and civil liberties, further disempower marginalized groups, and increase the role that private AI companies play in government decision-making. The way modern automated decision-making systems arrive at decisions can at times be impossible to fully understand even for the data scientists who designed them, in part because these systems might use complex models with thousands of variables. Since AI and ML can yield results that are counterintuitive with flaws that might be challenging to detect, many argue that the results and decisions are not accountable.

## THE US DHS AI-SURVEILLANCE PROGRAM

Though many of the reasons for incorporating AI into their border enforcement technologies are similar to Canada's, the use of AI and ML software differs immensely in practice. While Canada uses AI and ML for asylum processing of applications, DHS agencies like CBP and ICE use AI and ML to analyse data acquired through their robust surveillance program. Furthermore, the US asylum system is set up very differently to Canada's asylum process. For humanitarian migrants at the border, applying for asylum is much more complex. One of the most common methods used for humanitarian migrants upon a port of entry into the US is to declare they intend to seek asylum or fear persecution. Once they have legally declared to the CBP agent or ICE officer, migrants must then pass through an initial "credible fear" interview with CBP agents, which the agent will use to determine whether or not they are eligible to apply for asylum. During the "credible fear" interview, migrants must prove their fear and reason for seeking refuge is valid, while the reviewing officer reviews their case and any data or information they may already have. If the migrant passes the initial interview at the border, they must then apply for asylum which, due to a backlog of asylum cases, may take several years for a case to be heard. Over recent years, asylum policies have changed which make eligibility for granting asylum cases much more difficult than in the past.

The emergence of the official CBP drone surveillance program in 2004 resulted in a new way in which migration, specifically asylum, has come to be enforced under the DHS (Michel, 2015). The focus on enhanced security at the border by the US is a relatively novel concept, which only originated after the September 11, 2001 attacks, and under the Homeland Security Act of 2002, which was the establishment of the DHS as an executive department to prevent terrorism and minimize threats to the US (US Department of Homeland Security, 2002). Moreover, CBP and ICE agencies were created under the DHS to prevent any threats at the border. Since DHS was enacted, the department in total has spent over $US44 billion in securing the border through new AI surveillance technologies (Mijente, 2018). The predecessor to the current CBP surveillance program in operation, SBI-Net, was the first attempt to "develop a single technology that could be used across the border" (Preston, 2019). After nearly US$1 billion spent on SBI-Net, however, former Homeland Security Secretary, Janet Napolitano, cancelled the program and instead sought to implement a wider range of newer and more complex technologies through companies such as Boeing, Lockheed Martin, and Palantir (Preston, 2019). The incorporation of private defense contractors and tech firm giants have led to a privatized border. Both SBI-Net and the current CBP drone surveillance programs are excessively costly in the government's aim to provide a more accurate and neutral analysis in the risk assessment of migrants at the border.

The current drone program in operation was further implemented as a means to cut down the risk of and dependency on human agents in the field, while the deployment of military-grade surveillance drones and software were thought to increase border apprehensions. CBP uses a combination of larger, military-style drones and smaller hand-launched devices from a range of different manufacturers such as Lockheed Martin and Aeryon Labs (Ghaffary, 2019). The most notorious surveillance drone is perhaps the Predator B aircraft (Ghaffary, 2019). Its predecessor, the “A,” was used extensively by the US Military, beginning its service as a surveillance tool in the Balkans intervention in the 1990s before it “evolved [into] an armed flying weapon” (Thomson, 2017). Since their implementation under the DHS, drones have used a range of technologies to help the CBP and ICE surveil and monitor suspicious activity on and around the border. While the Predator B can stay in the air for up to 30 hours, its capabilities are increasingly outdated compared to the newer, smaller devices being used by the US. The Predator can read a number plate from roughly two miles in the air and not only requires far less training to pilot, but also contains ML algorithms that can identify human activity through its patchwork of sensors (Ghaffary, 2019). Additionally, drones have advanced to incorporate facial recognition technology, live-feed video cameras, thermal imaging, fake cell phone towers to intercept calls and extract GPS coordinates (Benson, 2015).

AI surveillance as a smart border technology continues to accelerate in infrastructure under the guise of the need to enhance security measures at the border due to perceived threats or crises. Incidentally, the increased insecurity of threats as well as actual threats have created a “security-insecurity paradox” (Chebel d’Appollonia, 2012: 2). Immigration policies coupled with new AI surveillance technology have made the asylum process severely restrictive in who can apply and who is then granted asylum. Together, the smart technology infrastructure and newer policies have transformed to fight terrorism while simultaneously managing the flow of asylum seekers and minority groups through counterterrorist surveillance policies (Chebel d’Appollonia, 2012: 2). Extensive amounts of both public and private data on migrants through AI surveillance are stored and collected for up to five years and can be shared amongst agencies (US Department of Homeland Security, 2018). The ability to share or spread highly sensitive data on migrants between intelligence and enforcement agencies increases the risk of mismanaged data. The Inspector General’s audit report found that the data collected is then used during encounters or interviews with CBP for further risk assessment and processing decisions (US Department of Homeland Security, 2018). Moreover, the use of data collected is processed by AI algorithms that are commonly used to identify and deter people who are likely to or have already committed crimes, thus pushing policies and practices that criminalize vulnerable migrants who are seeking refuge or asylum at the border.

Consequently, the security-insecurity paradox which continues to justify the need for AI surveillance to detect threats along the US-Mexico border is used to manufacture crises. In 2018, the migration caravans and family separation turned into a humanitarian "crisis" at the border because of the smart border technologies used by the CBP and ICE. Despite the notion that technology is objective, unequal distributions of power are implicit in smart border technologies and remain a clear and direct violation of humanitarian migrants' rights to privacy and protection. In the spring of 2018, as thousands of Central American migrants gathered in their journey to reach the border and claim asylum in the US, extensive CBP surveillance monitoring practices targeted many of these migrants. As the migrant caravans in 2018 became a looming threat towards US immigration enforcement, US Military troops were brought down to help secure the border under President Trump's declaration of a national emergency (Rohrlich, 2020). While most migrants remain relatively unaware of how much data and information is being collected on them, current surveillance technology breaks the ethics of consent and remains a gross violation of privacy that is used to manage migration flows and deter migrants from seeking asylum in the US.

Digital infrastructure has become just as important, if not more, to the physical border infrastructure (Gillespie et al., 2016: 2). However, there is major concern over accuracy and reliability of AI and case management software that stirs many trepidations over the exclusion of ethics and civil liberties (Dave, 2017). As many of the nuanced AI and surveillance technologies are part of highly confidential pilot programs, they remain testing grounds for the government and private companies alike. An additional part of what had led to the humanitarian crisis at the border was the introduction of the zero-tolerance pilot program, prior to having become an official policy under the Trump administration (Seville and Rappleye, 2018). This program, which was introduced to DHS in September of 2017, is highly controversial as it laid the blueprint for massive human rights abuses by the government in forcibly separating families. As a pilot program, a code of ethics remained absent from the design logic. This, perhaps, could be due to the overlooked assumption that technology is objective and therefore more ethical in managing asylum. By relying on such notions of objectivity, discriminatory practices of risk-assessment and biased AI-backed algorithms programmed to target individuals are an abhorrent violation of human rights.

Further general concerns over AI surveillance technologies are the well-known race and gender biases that exist within most AI surveillance algorithms. Samuel (2019) argues that AI biases are notoriously known for disproportionately targeting people of color. This is especially concerning, seeing as many of the Central American migrants fleeing persecution at the border are people of color. Despite the discriminatory practices of risk assessment and biased AI-backed algorithms, the DHS continues to operate its pilot programs through the rhetorical framework that "sacrificing the rights of the 'others' will protect 'us'" (Chebel d'Appollonia, 2012: 10). Such framework is the driving force behind the current changes in practices and policies of asylum that reinforce the government's push for increasing AI surveillance technologies. During the zero-tolerance pilot program, migrants who declared their intent to apply for asylum at the border have been detained in bleak and overcrowded facilities for an indefinite amount of time. More concerning was that while migrants awaited their "credible fear" interview in detention centers, CBP forcibly separated roughly 2,700 children due to the zero-tolerance pilot operation flagging anyone as a security risk, with no system put in place to reunite families (Lind, 2018). Despite the Flores settlement of 1997, which states families must be released from detention within 20 days, the current Trump administration has defied this ruling by keeping certain families or specific family members for an indefinite amount of time (Lind, 2018). Moreover, the stripping of migrants' rights and agency is used to intimidate and deter humanitarian migrants seeking refuge through the support of AI surveillance and risk analysis software.

As DHS surveillance programs become more prominent in enforcing asylum at the border, many humanitarian migrants are using technology to fight back from the government in an attempt to sustain their agency. One way in which migrants have fought back against border security measures, in addition to social media, is by using mobile networks to take different paths of migration in an effort to remain undetected from surveillance drones (Ghaffary, 2019). However, the use of technology by migrants along the border is more of a method of survival against the smart border technologies used by DHS authorities. Moreover, the access to smart phones and digital migratory networks is found to be both "gendered and generational" (Gillespie et al., 2016: 9), causing a rift in disparity levels amongst humanitarian migrants on their journeys to the border. Newer paths taken to avoid fiber-optic sensors and geolocation networks are much riskier and have thus resulted in a higher mortality rate of migrants (Ghaffary, 2019). The advanced capabilities of the surveillance drones used beyond the border are not only, for the most part, undetectable by migrants, but also versatile in that such technology can link together data from multiple sources (Fussell, 2019). As geolocation networks and fiber-optic sensors are used as an experimentation against humanitarian migrants to manipulate migratory paths, the increased deaths along the border and also in detention facilities defy almost every principle of the Nuremburg Code of Ethics, which set out to protect human rights when used for experimentation. Such advanced capabilities in the detection of suspicious activity of migrants has not only led migrants to take riskier paths in attempting to escape capture or being apprehended at the border, but may also be a reason why more migrants pursue illegal entry into the country.

The current way in which the DHS enforces immigration through AI surveillance technologies is concerning as the consequences of targeting and profiling humanitarian migrants leads to a manufactured "crisis" which is used to justify the need to impose stricter immigration regulations. While the asylum system is backlogged, detaining migrants for an extended amount of time before appearing for a credible fear interview allows CBP to gather as much evidence against migrants, through both public and private data. Such practices continue to alter the asylum system and modify such policies that were originally put in place to help humanitarian migrants into a way of criminalizing migrants. Thus, AI technologies have allowed for heavy discriminatory practices on and beyond the border, making it exceedingly difficult for humanitarian migrants to apply for asylum.

## CONCLUSION

Although there has been an increasing amount of attention in the public and private spheres on the uses of AI-powered technologies, the full capabilities of AI are still largely unknown. As a result, when implemented into immigration practices, AI remains highly experimental, costing humanitarian migrants their right to privacy and protection, consequently leading to discriminatory policies that target migrants and strip them of their agency.

While both Canada and the US have recently incorporated AI and ML into their immigration enforcement technologies, their implementation varies in practice. In the US, AI is predominantly used to identify and detect migrants who may pose a level of threat or risk along the border, while Canada uses AI and ML to generate automated decisions upon applications. Further expansions of the use of AI in the Canadian and US immigration spheres are on the way, including decisions on whether an applicant will be designated as a risk to Canada. Perhaps the reason for their different implications is that Canada's border remains relatively invisible, especially when compared to the US-Mexico border. Continued unregulated use of AI and ML by Canadian and US federal governments will make immigration and refugee law an immoral experiment that will increase discrimination towards highly vulnerable populations, and reduce the amount of accountability the Canadian and US federal government has towards the refugee regime. Paradoxically, humanitarian migrants fleeing their country for fear of persecution are ultimately stripped of even more of their human rights when met with unethical AI-based immigration enforcement practices.

As countries around the world, including the US and Canada, continue to increase their reliance on AI and ML, we have several recommendations. Increase the level of collaboration between key government agencies and stakeholders, as well as academics and civilians, so as to better understand the current and potential impacts of AI and related technologies on human rights. Publicize the source codes developed internally and externally for all federal government decisions that are made by AI and automated decision systems, except when such publication would risk privacy and national security. Develop a system for discerning which types of administrative processes are appropriate for use in automated decision technologies and border enforcement and which are not. Create a body that will oversee the use of automated decision systems by a third-party ethics committee to maintain checks and balances. This body must be impartial and at arm's length and be allowed to oversee all aspects of the system, as well as test and audit source codes. Furthermore, establish government-wide standards for the use of AI and automated decision systems, and encourage all government bodies using AI to publish reports revealing how they use them. Forbid the development and provision of any new AI and automated decision systems until the existing system is fully compliant with the above. By fostering a culture of interparty dialogue and openness, these technologies can either be restricted in the future or used in manners that benefit those most vulnerable. The data and the ways in which it is collected are essential if they are to be just in nature.

# REFERENCES

Bansak, K., Ferwerda, J., Hainmueller, J. et al. 2018. Improving Refugee Integration through Data-driven Algorithmic Assignment. *Science*, 359(6373): 325–9.

Beduschi, A. 2018. The Big Data of International Migration: Opportunities and Challenges for States under International Human Rights Law. *Georgetown Journal of International Law*, 49(2): 982–1017.

Benson, T. 2015. 5 Ways We Must Regulate Drones at the US Border. Wired.com. Condé Nast. Available at: https://www.wired.com/2015/05/drones-at-the-border/. Accessed June 19, 2020.

Bloomberg Politics. 2019. Palantir CEO Karp on Silicon Valley, ICE, 2020 Election. Available at: https://www.youtube.com/watch?v=1zHUXGd4gJU. Accessed June 20, 2020.

Braga, M. 2017. A Secretive Silicon Valley Tech Giant Set up Shop in Canada. But What Does It Do? CBC.ca. *CBC/Radio-Canada*. Available at: https://www.cbc.ca/news/technology/palantir-silicon-valley-technology-giant-data-canada-1.4111163/. Accessed June 1, 2020.

Calo, R. 2017. Artificial Intelligence Policy: A Primer and Roadmap. *UC Davis Law Review*, 51: 399–435.

Chebel d'Apollonia, A. 2012. *Frontiers of Fear: Immigration and Insecurity in the United States and Europe*. Ithaca, NY: Cornell University Press.

DarkMatter. 2016. DarkMatter Inaugurates R&D Centre Based in Toronto, Canada. *CNW Group Ltd.* Available at: https://www.newswire.ca/news-releases/darkmatter-inaugurates-rd-centre-based-in-toronto-canada-575528221.html. Accessed 4 June 2020.

Dave, A. 2017. Digital Humanitarians: How Big Data Is Changing the Face of Humanitarian Response. *Journal of Bioethical Inquiry*, 14(4): 567–9.

Flach, P. 2012. *Machine Learning. The Art and Science of Algorithms That Make Sense of Data*. Cambridge: Cambridge University Press.

Foucault, M. 1977–78. *Security, Territory, Population: Lectures at the Collège de France 1977–1978*, pp. 1–4; see notes 1–4 on p. 24 in Foucault (2007).

Foucault, M. 2007. *Security, Territory, Population: Lectures at the Collège de France, 1977–78*. Graham Burchall (trans.). London: Palgrave Macmillan.

Franco, M. 2019. Democrats Want a "Smart Wall". That's Trump's Wall by Another Name. Available at: https://www.theguardian.com/commentisfree/2019/feb/14/democrats-wall-border-trump-security. Accessed June 16, 2020.

Fussell, S. 2019. The Endless Aerial Surveillance of the Border. *Theatlantic.com*. Available at: http://www.theatlantic.com/technology/archive/2019/10/increase-drones-used-border-surveillance/599077/. Accessed June 17, 2020.

Ghaffary, S. 2019. The "Smarter" Wall: How Drones, Sensors, and AI Are Patrolling the Border. *Vox*. Available at: http://www.vox.com/recode/2019/5/16/18511583/smart- border-wall-drones-sensors-ai. Accessed June 15, 2020.

Gillespie, M., Ampofo, L., Cheesman, M., et al. 2016. Mapping Refugee Media Journey. *The Open University*. France Médias Monde. Available at: https://www.open.ac.uk/ccig/sites/www.open.ac.uk.ccig/files/Mapping%20Refugee

%20Media%20Journeys%2016%20May%20FIN%20MG_0.pdf. Accessed January 2021.

Government of Canada. 2019. How Canada's Refugee System Works. Available at: https://www.canada.ca/en/immigration-refugees-citizenship/services/refugees/canada-role.html. Accessed June 2, 2020.

Greene, T. 2019a. Study: Trump's Paid Peter Thiel's Palantir $1.5B So Far to Build ICE's Mass-surveillance Network. *The Next Web*. Available at: https://thenextweb.com/artificial-intelligence/2019/08/12/study-trumps-paid-peter-thiels-palantir-1-5b-so-far-to-build-ices-mass-surveillance-network/. Accessed June 8, 2020.

Greene, T. 2019b. Report: Palantir Took over Project Maven, the Military AI Program Too Unethical for Google. *The Next Web*. Available at: https://thenextweb.com/artificial-intelligence/2019/12/11/report-palantir-took-over-project-maven-the-military-ai-program-too-unethical-for-google/. Accessed May 8, 2020.

Hemmadi, M. 2019. Controversial Data-mining Firm Palantir Signs Million-dollar Deal with Defense Department. *The Logic Inc*. Available at: https://thelogic.co/news/exclusive/controversial-data-mining-firm-palantir-signs-million-dollar-deal-with-defence-department/. Accessed June 1, 2020.

Jacobsen, K.L. 2016. *The Politics of Humanitarian Technology: Good Intentions, Unintended Consequences and Insecurity*. London: Routledge/Taylor & Francis Group.

Keung, N. 2017. Canadian Immigration Applications Could Soon Be Assessed by Computers. *Toronto Star*. Available at: https://www.thestar.com/news/immigration/2017/01/05/immigration-applications-could-soon-be-assessed-by-computers.html. Accessed June 16, 2020.

Khandaker, T. 2018. Canada Is Using Ancestry DNA Websites to Help It Deport People. *Vice News*. Available at: https://news.vice.com/en_ca/article/wjkxmy/canada-is-using-ancestry-dna-websites-to-help-it-deport-people. Accessed June 2, 2020.

Lee-Shanok, P. 2019. Privacy Advocates Sound Warning on Toronto Police Use of Facial Recognition Technology. *CBC/Radio-Canada*. Available at: https://www.cbc.ca/news/canada/toronto/privacy-civil-rights-concern-about-toronto-police-use-of-facial-recognition-1.5156581. Accessed June 10, 2020.

Lind, 2018. The Trump Administration's Separation of Families at the Border, Explained. *Vox*. Available at: https://www.vox.com/2018/6/11/17443198/children-immigrant-families-separated-parents. Accessed June 2, 2020.

MacMillan, D. and Dwoskin, E. 2019. The War Inside Palantir: Data-mining Firm's Ties to ICE under Attack by Employees. *The Washington Post*. Available at: https://www.washingtonpost.com/business/2019/08/22/war-inside-palantir-data-mining-firms-ties-ice-under-attack-by-employees/. Accessed June 11, 2020.

Mazzetti, M., Goldman, A., Bergman, R. and Perlroth, N. 2019. A New Age of Warfare: How Internet Mercenaries Do Battle for Authoritarian Governments. *The New York Times*. Available at: https://www.nytimes.com/2019/03/21/us/politics/government-hackers-nso-darkmatter.html. Accessed January 2021.

Michel, H.M. 2015. Customs and Border Protection Drones. *Drone Center*. Available at: https://dronecenter.bard.edu/customs-and-border-protection-drones/. Accessed June 8, 2020.

Mijente. 2018. Who's Behind ICE? *National Immigration Project*. Available at: https://www.nationalimmigrationproject.org/PDFs/community/2018_23Oct_whos-behind-ice.pdf. Accessed June 20, 2020.

Molnar, P. and Gill, L. 2018. Bots at the Gate: A Human Rights Analysis of Automated Decision-making in Canada's Immigration and Refugee System. *International Human Rights Program and the Citizen Lab*. Available at: https://citizenlab.ca/wp-content/uploads/2018/09/IHRP-Automated-Systems-Report-Web-V2.pdf. Accessed June 10, 2020.

Ng, V. 2017. Algorithmic Decision-making and Human Rights. The Human Rights, Big Data, and Technology Project. *Essex University's Human Rights Centre*. Available at: https://www.hrbdt.ac.uk/algorithmic-decision-making-and-human-rights/. Accessed June 7, 2020.

Peritz, I. and Leblanc, D. 2017. RCMP Accused of Racial Profiling over "Interview Guide" Targeting Muslim Border Crossers. *The Globe and Mail*. Available at: https://www.theglobeandmail.com/news/national/rcmp-halts-use-of-screening-questionnaire-aimed-at-muslim-asylum-seekers/article36560918/. Accessed June 10, 2020.

Peterson, B. 2020. Insiders Say a Palantir Exec Claimed Profitability and Compared Project Maven to the Nuclear Bomb in January All-hands. Available at: https://www.businessinsider.com/palantir-executive-compared-project-maven-to-manhattan-project-2020-1?r=US&IR=T. Accessed June 10, 2020.

Preston, D. 2019. An Ancient City Emerges in a Remote Rain Forest. *The New Yorker*. Available at: http://www.newyorker.com/tech/annals-of-technology/an-ancient-city-emerges-in-a-remote-rain-forest. Accessed June 10, 2020.

Rango, M. 2015. How Big Data Can Help Migrants. *World Economic Forum*. Available at: https://www.weforum.org/agenda/2015/10/how-big-data-can-help-migrants. Accessed June 2, 2020.

Rohrlich, J. 2020. Court Document shows US Troops Surveilling Migrants at the Mexico Border. *Quartz*. Available at: https://qz.com/1815249/us-troops-are-surveilling-migrants-along-the-border-with-mexico/. Accessed May 15, 2020.

Samuel, S. 2019. Some AI Just Shouldn't Exist. *Vox*. Available at: https://www.vox.com/future-perfect/2019/4/19/18412674/ai-bias-facial-recognition-black-gay-transgender. Accessed May 15, 2020.

Satzewich, V. 2014. Canadian Visa Officers and the Social Construction of "Real" Spousal Relationships. *Canadian Review of Sociology*, 51(1): 1–21.

Seville, R. and Rappleye, R. 2018. Trump Admin Ran "Pilot Program" for Separating Migrant Families in 2017. *NBC Universal*. Available at: https://www.nbcnews.com/storyline/immigration-border-crisis/trump-admin-ran-pilot-program-separating-migrant-families-2017-n887616. Accessed June 16, 2020.

Shepherd, M. 2017. RCMP Will Redact More Than 5,000 Records Collected Using Questionnaire Targeting Muslim Asylum Seekers. *Toronto Star*. Available at: https://www.thestar.com/news/canada/2017/11/27/rcmp-will

-redact-more-than-5000-records-collected-using-questionnaire-targeting-muslim-asylum-seekers.html. Accessed June 2, 2020.

Thomson, I. 2017. US Air Force Terminates Predator Drones. Now You Will Fear the Reaper. Available at: https://www.theregister.com/2017/02/27/us_air_force_put_predator_drones/. Accessed January 2021.

UKGOS. 2016. Artificial Intelligence: Opportunities and Implications for the Future of Decision Making. *The Government of the United Kingdom.* Available at: https://assets.publishing.service.gov.uk/government/uploads/system/uploads/attachment_data/file/566075/gs-16-19-artificial-intelligence-ai-report.pdf. Accessed June 1, 2020.

US Department of Homeland Security. 2002. *Homeland Security Act 2002*, Public Law 107-296. 107th Congress Act. Available at: https://www.dhs.gov/sites/default/files/publications/hr_5005_enr.pdf. Accessed June 3, 2020.

US Department of Homeland Security, Office of Inspector General. 2018. OIG-15-17 – US Customs and Border Protection's Unmanned Aircraft System Program Does Not Achieve Intended Re. 2014.

Wall, M. 2020. Information Precarity. In K. Smets, K. Leurs, M. Georgiou, S. Witteborn and R. Gajjala (eds), *The Sage Handbook of Media and Migration.* London: Sage, pp. 85–90.

# 5. Migration and smuggling across virtual borders: a European Union case study of internet governance and immigration politics

**Johanna Bankston**

## INTRODUCTION

The advent of social media in the early 2000s facilitated fast and easy communication between anonymous users regarding illicit activities, making social media communication a Pandora's box that governments worldwide were unprepared to handle. One issue that governments must contend with is the planning, solicitation, and facilitation of migrant smuggling that occurs on social media, other internet services providers (ISP), and information communication technologies (ICT). Social media in particular has allowed smugglers to profit immensely from the rising interest in asylum and economic migration to the global north from many groups in the global south. However, social media has also made smugglers' activities more difficult to detect and prosecute. Member states of the European Union (EU) have responded to this online trend by implementing ad hoc efforts to police content related to migrant smuggling on social media and by pushing legislation that would make social media companies more responsible, through liability or sanctions, for illegal content generated by their users. These state-led efforts to deter illegal migration through the policing of information communication raise important concerns about the scope of EU internet governance; what bearing does internet governance have on migration politics? How is migrant empowerment associated with access to social media and to which risks will changes in internet governance structures subject them?

This chapter explores the above questions by analysing migration politics online through the concept of internet governance at the micro and macro levels (Brooks 2015: 13, 18). I use this conceptual framework to explore how social media companies and EU governments understand and negotiate their responsibilities towards online smuggling activity and migrant user privacy. I consider how these negotiations and digital policies may place limitations on migrant agency.

The chapter is divided into four sections. The first section addresses micro-level internet governance of social media spaces by smugglers and migrant users. In the second section, I consider EU commitments to data privacy, to migrant safety and smuggling prevention, and their formal and informal actions to remove smuggling related content on social media at the macro level of internet governance. The third section offers an analysis of one social media company's responses and official policies towards smugglers using their platforms. In the final section, I argue that EU member states engage in internet governance to deter irregular migration through 'virtual border enforcement.' I also argue that the push to increase ISP liability represents state coercion of non-state actors to perform border enforcement in sociotechnical spaces and I highlight the transformation of content moderation into a technology of border enforcement. I conclude by emphasizing the dangers of creating digital policies and partnerships aimed at subverting human smuggling which would eclipse migrants' digital rights and identities.

## MIGRANTS ON SOCIAL MEDIA: AGENCY AND RISK

### Micro-level Internet Governance

Migrants increasingly engage with social media to inform their migration plans, to navigate their journeys, and to integrate themselves into their new homes post-migration. This engagement is shaped by many factors, including the design of the platform on which the migrant uploads content, the laws that regulate how an internet service may be used, and also by the knowledge produced by the content of other users. These factors interact in a process of 'Internet Governance,' which is generally defined as 'the development and application by governments, the private sector, and civil society, in their respective roles, of shared principles, norms, rules, decision-making procedures, and programs that shape the evolution and use of the Internet' (Brooks 2015: 13). Migrants who use the platforms to plan, navigate journeys, and ease their integration contribute to internet governance as members of civil society at the micro level or the user level (ibid.: 18). For the purposes of this chapter, I employ the term 'migrant users' to refer to individual users who either claim to be current, former, or potential migrants or individual users who engage with social media to facilitate their migration. While the term 'migrant users' is an umbrella term that covers users interested in business, student, family reunification and many other kinds of immigration, this chapter focuses on irregular migrants interested in asylum or economic migration who make their journeys without pre-arranged permission to enter their desired country of destination in the EU.

Through their user-generated content, actions, and interactions with other users, migrant users create norms regarding the irregular migrant experience and its intersection with technology. The creation and sharing of content regarding their journeys to the EU produces aggregated knowledge about the importance of certain factors at the pre-departure, in-transit, and post-arrival stages of migration. Their actions and narratives produce and spread information about which authorities are trustworthy, which governments and locations are desirable as permanent homes, whether or not the benefits of migration outweigh the costs, and who is most suited to succeed or fail as an immigrant living abroad, among other narratives. Their very presence on social media platforms as 'migrant users,' and their willingness to post about migration related questions, concerns, and advice in public internet spaces, establish a norm that migrant identity, solidarity, and mobilization are legitimate in sociotechnical spaces. Migrant voices online also create knowledge about the physical objects, terrain, and conditions associated with migration through the uploading and dissemination of pictures, videos, testimony, advice, and other evidence of their movement across international borders (Frouws et al. 2016: 3). Such proofs may be used to reassure family members and communities of their safety or to instruct other migrants about potential dangers or methods of safe passage (ibid.: 2–3). However, their voices in sociotechnical spaces may also be viewed and used by unintended audiences, such as non-governmental organizations (NGO) interested in migration patterns or states interested in using social media as a surveillance technology to limit migrant access to their borders (ibid.: 12).

I emphasize here that the production of norms and truths by migrant users regarding the migrant experience is not wholly representative. The demographics of usership (gender, age, religion, sexual orientation, ethnicity, and so on) and the different levels of engagement of users foreground some migrant experiences while hiding or dismissing others.

For example, research from the Harvard Humanitarian Initiative found that female migrants in their study were 27 percent less likely to have access to mobile phones and internet connectivity than men (Latonero et al. 2018: 5). The implication of this unequal access is that production of knowledge about migrants through content online is often skewed to reflect the experiences of migrants with the most access and tech literacy, and is not representative of all migrants. The effects this phenomenon has on the constitution of digital identities and communities is an area for further research.

## Migrant Digital Identity and Agency

Many migrants claim that access to a smartphone is vital to the success of their journey (Latonero et al. 2018: 3). For migrants with access to smartphones, social media is an increasingly important source of connectivity and information (Frouws and Brenner 2019). Migrants rely on social media for family connectivity, money transfers, job searches, and connection to networks of other migrants or diasporas (Alencar et al. 2018: 835). Access to social media can also minimize risks of death or disappearance, as migrants can speak with family members or other contacts in emergency situations such as kidnapping, extortion, arrest, or medical emergency. There is potential for social media to promote the safe passage of migrants by connecting users, disseminating informative content, and giving migrants access to the resources they need to navigate their journeys independently or in a migrant-led group. The creation of migrant solidarity groups on platforms like Facebook, with names like 'Europe Without Smugglers', enables migrants to crowdsource information from others who have completed their travel routes about road blocks, police or military checkpoints, and other dangers associated with certain routes (Frouws et al. 2016: 5).

Many migrants still choose to hire smugglers to help them reach their desired destination due to the high levels of risk and uncertainty involved in travel (Alencar et al. 2018: 837). Using a smuggler takes pressure off migrants to plan the logistics of migrating through various countries, or to anticipate all of the obstacles in their journey, a task which is extremely challenging for many migrants who may have limited knowledge of the countries they are passing through. Social media may enable migrants to minimize their risk of exploitation or harm from smugglers by linking them into what the European Commission referred to in a 2016 EMN Inform report as a virtual 'marketplace' of smuggling services (European Commission 2016a: 3). In this virtual marketplace, services, prices, routes, and expectations surrounding the migrant and smuggler relationship are standardized to some extent through a process of advertisement, testimony, and reviews. Migrants can give negative reviews of smugglers by commenting on their advertisements or warning other migrants on related pages, and migrants can obviate certain risks by researching routes ahead of time or asking former migrants about their experiences (Roberts 2017: 28).

The spreading of information on the dangers associated with smuggling, including the exposure of lies or abuse by smugglers, is an important aspect of knowledge production on the migrant experience at the micro level. The spread of this information is a form of 'counter narrative,' or the spreading of information that runs counter to disinformation online (European Commission 2015: 6). Counter narrative by migrant users can be an efficient method of user-led regulation in which migrant users decide which smugglers will have success finding clients online and what the standards for the smuggler-client relationship will be.

Governments and NGOs also employ counter narratives to spread information about the dangers of smuggling online. However, research from the Mixed Migration Center shows that information from government and NGO sources are not usually successful in reaching migrants, as most tend to turn to other migrants or family and friends to inform their decisions about smugglers (Frouws and Brenner 2019). First-hand counter narratives by migrants themselves about the risks involved with smuggling, or with certain smugglers, on social media platforms are an important source of information that has great potential to promote safe and legal passage to Europe. While these methods of protection are imperfect and limited, as migrants will still run risks of harm whether or not they take a self-guided journey or a smuggler-guided journey to the EU border, they are significant examples of migrant agency to protect and empower themselves and other migrants through user-generated content and online interactions. It also represents how migrant user activity dictates norms about the use and utility of social media platforms at the micro level of internet governance.

While the use of social media gives migrants a significant advantage in the planning and navigating of their journeys, it may also pose some risks to migrant users seeking to enter the EU, North America, or other territories with Non-Entree regimes in place. The first is the risk of contracting a smuggler through social media who may violate the migrants' human rights and put their life in danger during transit. The second risk is that governments may use digital policy to police social media to undermine the agency of migrant users in the name of smuggling deterrence.

## Smuggler Agency Related to Social Media

Smugglers are key players in the micro-level governance of sociotechnical space, as they also contribute to the practice of norm setting and knowledge production of migrant journeys. In the past two decades, human smugglers have used new technologies to make themselves both more visible and accessible to migrants and, at the same time, more difficult for governments to track and prevent (Frouws and Brenner 2019). In addition to dual anonymity (from governments and authorities) and expansive, low-risk exposure (to potential migrant clients), social media platforms offer smugglers the following operational advantages:

1. Facilitation of operations from outside of the EU minimizes risk of capture or imprisonment by EU authorities.
2. Access to greater numbers of potential clients in a virtual 'migration marketplace.'
3. Large degree of control over the narrative and image of services being offered.
4. Greater ease and efficiency in communications with migrants/facilitators during transit.

The designs of social media sites, such as Facebook, enable users to publicly advertise their businesses or pages to potential buyers or followers using a variety of tools. Smugglers advertise their services on open and closed groups, discussion posts, individual profiles, ad campaigns, and pages (Roberts 2017: 13–14). Facebook features allow smugglers to target audiences by age, region, language, gender, education, and related interests, among other categories, allowing smugglers to identify audiences rapidly and with greater accuracy. Smuggling services advertised on Facebook typically host a range of options including genuine (stolen or borrowed) passports of EU citizens, fake passports, forged or real EU birth certificates, assistance with tourist visa applications, fake visa paperwork, travel plans via airplane, boat, automobile, cargo ship, and cargo truck (Adamson and Akbiek 2015). Some advertisements drive traffic by being rather blatant about the nature of their operation by detailing dates, departure locations, contact details, benefits in the destination country, prices by service, routes, supplies, risks involved, and plans to use falsified documents (ibid.). BBC news translated the following Facebook post:

> Travelling straight to Europe by plane costs between 4,000 and 8,000 euros. If you only want a passport and pay for the plane ticket yourself, and you take the consequences, then the price is 4,000 euro for a 100% genuine passport belonging to someone who looks like you. But if you want to pay after arriving in Europe, the price is 8,000 euro and we will pay for everything and take all the risk. Just send your picture and your number via Facebook message … (Adamson and Akbiek 2015)

Posts of the above nature run a higher risk of being detected or flagged quickly by authorities, users, or ISP content moderators. Zoe Roberts argues that this kind of cavalier behavior may demonstrate that risks associated with detection or content removal are sufficiently low for smugglers, and the rewards associated with more explicit posts are sufficiently high (Roberts 2017: 21). While it is relatively easy for ISPs to identify and take down these types of posts, other smuggling service pages evade detection for longer periods because they masquerade as legitimate travel agencies, making it difficult for ISP content task forces to identify them and verify the legitimacy of their operation. In 2015, an official Facebook advertisement campaign funded by a 'travel agency' offered a travel service from Libya to Europe through the Italian border (RT News 2016). This campaign likely surfaced in response to the passing of the EU-Turkey deal which aimed to deport migrants who arrived in the EU via Greece back to Turkey to await their immigration hearings. As a result of the deal, demand increased for alternative routes into Europe (ibid.). The campaign advertised safe passage for US$4,000 per person (ibid.). This campaign and similar posts included photos of yachts and luxury cruise ships, a tactic commonly used to mislead authorities about the nature of the operation, which may also inadvertently mislead migrants about the safety conditions of their travel (ibid.). Smugglers also evade detection by actions as simple as changing group names, deleting pages and creating new ones, or closing groups (Roberts 2017: 18). Further, Facebook has no requirements for identity verification, and fake profiles are an easy way for smugglers to give out their contact information without revealing their individual identity. EU data privacy laws allow ISPs to withhold the IP addresses of users from authorities, adding another layer of anonymity to the transaction (European Parliament and the Council of the European Union 2000; Kolodziejczyk 2015: 73–4). Once contact is established on social media, communication is transferred to private ICT platforms, making it difficult or impossible to track further, as many modern ICT platforms, such as Whatsapp, use end-to-end message encryption. Research by the Mixed Migration Center demonstrates that most migrants make their decisions to hire a smuggler whom they have met online only after speaking extensively with them on the phone or meeting them in person (Frouws and Brenner 2019). Their data also shows that, while social media is an increasingly popular way for migrants to source potential smugglers, most migrants prefer to meet smugglers locally and to make transactions in person whenever possible (ibid.).

It is important to state here that the relationship between smugglers and migrants is one of complexity and, at times, mutual benefit. EU states portray smugglers as unambiguous villains set on victimizing migrants. Such attitudes towards smugglers are reflected in the EU Plan of Action Against Migrant Smuggling, which describes smugglers as follows:

> Ruthless criminal networks organize the journeys of large numbers of migrants desperate to reach the EU. They make substantial gains while putting the migrants' lives at risk. To maximize their profits, smugglers often squeeze hundreds of migrants onto unseaworthy boats – including small inflatable boats or end-of-life cargo ships – or into trucks. Scores of migrants drown at sea, suffocate in containers or perish in deserts. (European Commission 2015: 1)

This rhetoric insinuates that all smugglers operate in organized criminal networks, that they are profit driven and have no personal investment in the safe passage of the migrants they smuggle, and it also explicitly blames them for migrant deaths at sea. While this characterization may represent some smugglers, Gabriella Sanchez's research has shown that many individuals who act as smugglers or help the facilitation of smuggling may be migrants themselves, individuals acting in solidarity with migrants, or they may be locals seeking to supplement their incomes due to economic strain (Sanchez 2017: 13). Sanchez's work also reveals that, contrary to state narratives of smugglers as organized criminals linked to human trafficking networks, historical records show that smuggling efforts are often independent and unorganized (ibid.: 14–15). EU policies do not reflect this reality because narratives of complexity do not serve border securitization goals.

However, while having access to a larger variety of smugglers and services may be an advantage for many migrants, and while some migrants do have positive experiences with their smugglers, research shows that contracting a smuggler is related to some risk of harm or death in transit (Mixed Migration Center 2018: 22). These risks disproportionately affect vulnerable migrants, such as women, children, and LGBTQ+ migrants (ibid.: 22). In addition to the risk of death to which some smugglers expose migrants, many report being threatened, sexually assaulted, physically abused, extorted, held captive, subjected to unsafe and unsanitary conditions, and in some cases being trafficked by the smugglers they hired (ibid.: 48). Such negative outcomes underscore the advantages of migrants' abilities to assess their transit options in a virtual marketplace based on information produced in-part by other migrants, but they also underscore the gravity of the dangers migrants face when some smugglers engage in knowledge production which is misleading or false. The nature of social media tools, as described above, makes it challenging for migrants to discern which users may be doctoring the images, services, and reviews related to their operation. While migrants are able to use social media to warn each other against certain smugglers, this is not an infallible method of protection, and migrants must grapple with the overlapping agency and risk that social media offers them in relation to informing themselves about options for safer passage.

## Social Media Use Post-arrival to the EU

Social media profiles may work to the advantage or disadvantage of migrants. Some migrants may be able to use their social media profiles to evidence their asylum claims, to aid their visa applications, or to contact friends and family to help facilitate the collection of evidence in support of their applications. However, governments may also police social media to gather information, and possibly evidence, against migrants seeking asylum or in deportation proceedings. The US Department of Homeland Security announced in late 2019 that they would begin requiring the social media account details (usernames and passwords) of visa and asylum applicants (Human Rights Watch 2019). Such measures have been called a deep invasion of privacy by Human Rights Watch (ibid.). It is less clear whether EU state authorities monitor migrant profiles in the same way that the US does but most EU governments do admit to policing social media for content related to human smuggling (European Commission 2016b). Further, through data-sharing agreements between the US and European governments, the social media information present in a migrant's US immigration file will also be accessible in the EU (Greenfield 2020). The potential of social media as a state surveillance tool for governments should not be underestimated, especially where migrants are concerned.

# EU ACTION: POLICIES, PRACTICES, AND INTERVENTIONS

## Macro-level Internet Governance by State Actors

While users generally create the norms of social media interaction over time at the micro level, their digital actions are enabled, influenced, and limited by the decisions of state and non-state actors at the macro level of internet governance. At the macro level, internet governance is negotiated by states and ISPs. Brandon Brooks describes macro-level internet governance as 'arrangements for how to distribute and coordinate Top-level Domains (TLDs) and data sharing agreements' (Brooks 2015: 18). Action and coordination at the macro level are not typically noticed by individual users at the micro level, but macro-governance continues to determine the architecture of the sociotechnical spaces in which users and organizations engage. For example, EU governments limit how and for what purposes social media companies can interact with user data through the General Data Protection Regulation (GDPR) (European Union 2016).

Social media companies contribute to internet governance by everything from the design of their platforms (the arrangement of how search results appear – by traffic, keyword matches, or purchase) to the requirements they set to create a user profile (whether or not a valid identity document, personal information, or an email is a requisite to create a user profile). These design decisions are political in nature, as they both define and restrict the capabilities of users and the constitution of digital identity. They also help determine the initial sociological construction of the public social media space (for example, offering certain identifiers such as age, marital status, and hometown but not others like political party affiliations or sexual orientation). It is therefore important to consider how changes to the architecture of these spaces through macro-level internet governance affects, and possibly undermines, the organization and mobilization of smugglers and/or migrant users at the micro level.

## EU Plan of Action Against Migrant Smuggling

Beginning in 2015, EU governments and authorities began to pay increased attention to the issue of smuggling on social media and the opportunity for deterrence that this activity presented to them. A 2016 ad hoc report by the European Commission inquired with states in the European economic area about their monitoring of social media for this activity (European Commission 2016b). Responses indicated large discrepancies in state approaches to social media surveillance. Fourteen out of the twenty-one respondents admitted to having police investigators and other relevant authorities monitor social media for smugglers through a variety of avenues, three said that they do not monitor social media for smuggling activity, and four declined to publish the answers they provided to the Commission (ibid.). Respondents' decisions about standards and conduct for monitoring social media activity were hardly uniform. For example, a number of states reported that they monitor open (public) groups, closed (private) groups, discussion pages, advertisement campaigns, and individual profiles while others reported that they only monitored public, open groups (ibid.). Governments who responded that they did not monitor social media websites for such activity at all indicated that they were constrained by data and internet privacy laws unique to their country and/or disparities in their resources or technological capabilities (ibid.). Few states responded positively to the question on whether social media evidence had been used (successfully or unsuccessfully) in criminal cases against human smugglers (ibid.). This points to another issue central to social media surveillance, that there is a general lack of standard legal practice for monitoring social media and for the use of electronic evidence. Another point that varied greatly was the respondents' respective commitments to partnerships with ISPs in combating content related to smuggling; at the time of the report, only seven out of sixteen EU member state respondents had formed such partnerships (ibid.).

The outcomes of subsequent meetings regarding migrant smuggling, including the EU Action Plan Against Migrant Smuggling and the EU Council Conclusions on Migrant Smuggling issued in March 2016, identified the following strategies for improved securitization in this area:

1. Standardize EU policy on social media monitoring and digital investigation.

2. Increase cooperation between governments and private social media companies.
3. Give greater responsibility to social media websites to detect smuggling content.
4. Spread counter narratives on social media to influence migrants to avoid smugglers.

A common theme among many of the ad hoc responses, also reflected in the subsequent action plan recommendations, was states' desire to see social media platforms take greater responsibility in the efficient flagging and removal of content (European Commission 2015). This may reflect recent changes in EU public opinion, among state officials and citizens alike, towards increased responsibility for ISPs resulting from a slew of data breaches and concerns over 'fake news' (Hoffman and Gasparotti 2020: 5). Indeed, in many EU government statements made to media sources with regards to this topic, there is a sense that government actors believe that the onus (ideologically and legally) should be on social media platforms to use their greater access and more advanced technological capabilities to tackle illegal content. At present, under EU e-commerce law, social media platforms are not liable for illegal content posted by their users though they are required to remove such material as soon it is brought to their attention by users, moderators, or authorities (Kolodziejczyk 2015: 61–2). Further, social media companies have no obligation to proactively monitor their users for illegal activity nor are they obligated to determine the standards for 'illegal' content at present (ibid.: 70).

## Digital Services Act

Legal obligations for social media companies operating in Europe may change in the near future when the long anticipated Pan-EU Digital Services Act (DSA) enters into force. In May 2020, the Committee on Civil Liberties, Justice and Home Affairs presented a draft report to the European Parliament which included a motion for a resolution on a new DSA, on the grounds that the now 20-year-old E-Commerce Directive fails to address much of the illegal content proliferated online today (Committee on Civil Liberties, Justice and Home Affairs 2020: 8). The new DSA would hold ISPs responsible, through liability or sanctions, for failures to remove illegal content swiftly and for lack of accountability and/or transparency in content moderation (ibid.: 5). The motion also calls for the creation of an independent EU regulatory body that would 'exercise effective oversight of compliance with the applicable rules; [and] . . . enforce procedural safeguards and transparency and provide quick and reliable guidance on contexts in which legal content is to be considered harmful' (ibid.: 6). The explanatory statement in the report suggests that the proposal is made in light of the drastic increase in illegal and harmful or offensive content online, which governments could not have anticipated when the E-Commerce Directive was created (ibid.: 8). While these concerns over data privacy and increased responsibility are legitimate, digital rights advocates indicate that there are legal and ethical concerns with the proposed DSA that should be addressed before its implementation. I consider these concerns as they relate to content concerning irregular migration, the right to free speech and expression, and the right to information online during a crisis (Greenwood et al. 2017: 13).

The issues of primary concern for migrant rights in the DSA are the motion for increased ISP liability and the motion to create a regulatory body to police 'legal, yet harmful' content (ibid.: 6). The former concern, raised frequently by digital rights activists, is that increasing liability for ISPs would inevitably result in overly cautious behavior by ISP content moderators. Not only would ISPs likely take down legal content that may be adjacent to or intersecting with illegal content, but it also might impede the development of more advanced and accurate moderation technologies, as there would be less need for them (Castro and Chivot 2019). For migrant users, this could mean the removal of important crowdsourced information about travel concerns, police checkpoints, high-risk travel routes, smugglers, and health and safety resources that could mean the difference between safe passage or death en route to the EU border. While the removal of content that is blatantly related to smuggling is often justifiable on legal grounds, risk of liability may cause ISPs to take down not only smuggling related content but all content related to irregular immigration, out of an abundance of caution. Further, because enforcement would be happening at the macro level, the policing and removal of migrant solidarity groups online would be largely invisible.

The latter concern, regarding the 'legal, yet harmful content' also has important bearings on migrant safety and freedom of expression online. The Center for Data Innovation (CDI) points out that the E-Commerce Directive requires companies to take down illicit information or activity, but the Directive does not include guidelines on the proper determination of illicit content or activity (Castro and Chivot 2019). At present, the

Court of Justice of the European Union (CJEU) has determined that ISPs are responsible to use 'automated search tools and technologies' available to them to identify equivalent examples of user-generated content that were previously found to be illegal when making decisions about the removal of flagged content (Committee on Civil Liberties, Justice and Home Affairs, Draft Report on the Digital Services Act 2020: 4). However, the Draft Report acknowledges that content removal decisions may be more difficult where context is required and that there is significant room for error to be made in this regard that would result in 'unnecessary restrictions being placed on the freedom of expression' (ibid.: 5). This refers to both illegal content and the 'legal, yet harmful' content over which the EU regulatory body would provide oversight. The draft refers to 'harmful and illegal content' throughout, justifying its proposal for the policing of 'legal, yet harmful' content by introducing examples such as the 'opaque political advertising and disinformation on COVID-19 causes and remedies' (ibid.: 4). Though these examples are significant, they obscure more controversial examples of what the EU may deem to be 'harmful' content. No context is given for how the regulatory body may determine the status of content as 'harmful,' however at point nine, the draft states that illegal content will be removed 'in line with European values' (ibid.: 5).

This rhetoric may have significant bearing on content related to irregular migration towards the European border, if border securitization and irregular migration prevention is deemed to be a value which dictates decisions on content removal. Legal content related to safe passage, asylum claims and proceedings, visa applications, detention and deportation, and related topics may be policed or removed because of their proximity or overlap with the topic of illegal border entry. An apt illustration of this would be the previously mentioned Facebook page 'Europe Without Smugglers.' Under the current E-Commerce Directive and the GDPR, Facebook allows this page to function so long as it does not explicitly reference or instruct others on illegal border crossing or other illegal behavior. The content on this and similar pages is useful for many migrants who may use it to protect themselves from harm or death in transit to European borders or to obtain important resources en route. The removal of legal content related to irregular migration experiences, advice, and questions would violate migrant users' rights to access information in a crisis (Greenwood et al. 2017: 13). If the proposed DSA enters into force, Facebook may choose to remove this and similar pages as they allude to the possibility of eventual illegal crossings of EU borders or they may be coerced by the EU regulatory body to remove this content for the purpose of deterring such attempts. While the DSA highlights the need to create content moderation regulation which is fair, equal, and does not unnecessarily infringe the right to free speech or expression, their 'legal, yet harmful' rhetoric raises important concerns about the potential for the EU regulatory body to abuse its discretion by using content moderation to filter out posts related to irregular migration.

## Protocol Against Migrant Smuggling

A fourth concern with digital monitoring is that EU governments will not only track and criminalize smugglers, but they will also track, deter, and attempt to criminalize migrant users. A common assumption is that governments may act in this way because migrants who hire smugglers are participating in criminal activity, and they therefore forfeit their rights. However, the Protocol on the Smuggling of Migrants sets forth rights and plans of action which aim to protect the rights of undocumented migrants while criminalizing smugglers.

> The Smuggling Protocol expressly says that the document is not intended to criminalise migrants, and that those migrating through smuggling routes have the right to life, the right not to be subjected to torture or other cruel, inhumane or degrading treatment or punishment, and the right to non-refoulement. States are further obligated to take 'appropriate measures' regarding protection from violence; assistance if a migrant's life or safety has been endangered; consular notification and communication; and safe and humane treatment during boat interdiction . . . Further, a broad savings clause obliges States not to implement the Protocol in a way that contravenes the human rights commitments they have made. (Global Alliance Against the Trafficking of Women 2011: 17)

Following the objects and purposes of the Protocol Against Smuggling, EU policy on the monitoring of public social media posts regarding smuggling content should identify criminal smugglers while respecting the anonymity and rights of migrant users. Article 10 item (f) of the Protocol states that EU members are empowered to mine and share information on 'Scientific and technological information useful to law enforcement, so as to enhance each other's ability to prevent, detect and investigate the conduct set forth in Article 6 of this Protocol and to prosecute those involved' (UN General Assembly 2000). The allotment provided by this clause should be supplemented by the creation of a standardized legal approach to social media monitoring for the purpose of tackling smuggling operations without criminalizing migrant users inquiring or speaking about irregular migration. This is a critical consideration for the DSA, as their guidelines on content moderation should take into account the rights of asylum seekers and other smuggled migrants.

## FACEBOOK: OFFICIAL POLICY, UNOFFICIAL PRACTICE, AND RESPONSIBILITY

### Smuggling Deterrence Policies and Content Moderation

While most social media platforms are explicit about their policies towards content which promotes human trafficking and sex-exploitation activity, they are less explicit about their responsibilities towards content related to human smuggling. Facebook states its official operating policies on its Community Standards page, which mentions four official, detailed policies on content and use related to human trafficking. However, the Community Standards page says only the following with regards to smuggling operations at Article 10:

> While we need to be careful not to conflate human trafficking and smuggling, the two can be related and exhibit overlap. The United Nations defines human smuggling as the procurement or facilitation of illegal entry into a state across international borders. Without necessity for coercion or force, it may still result in the exploitation of vulnerable individuals who are trying to leave their country of origin, often in pursuit of a better life. Human smuggling is a crime against a state, relying on movement, and human trafficking is a crime against a person, relying on exploitation. (Facebook Community Standards: accessed March 2020)

This justification reflects Facebook's position that the largest share of responsibility towards monitoring smuggling-related content lies with states. The only other mention of smuggling in Facebook's guidelines is a sentence which states that the platform shall remove any content which 'offers or assists in the smuggling of humans' (ibid.: accessed March 2020). Facebook hosts an annual Global Safety Summit to address pressing human rights and safety concerns that play out on its platform, including mental health, cyber-bullying, and terrorism, among others. The Summit regularly hosts a panel on human trafficking and sexual exploitation, but these panels do not typically address Facebook's actions towards content related to migrant smuggling. Internet searches yield no direct statements from Facebook regarding their actions to track or remove smuggling-related content on its platform, though the European Commission's Ad Hoc Query indicates that the company does cooperate with some EU member states and police authorities on this issue (European Commission 2016b). Facebook's language and lack of detailed address about migrant smuggling likely indicates that they are unable to give as much priority to monitoring this content as they do for human trafficking, child sexual exploitation, or other individual rights violations, which are more clearly defined legally. It may also indicate their caution to intervene in migrant smuggling-related content because it may result in unnecessary infringement on the rights of migrant users to self-expression and free speech, or the criminalization of people on the move who, as they write, are 'in pursuit of a better life' (Facebook Community Standards: accessed March 2020). Regardless, their silence on the issue leaves gaps in knowledge about Facebook's official tactics to detect and/or deter human smugglers on its platform in cooperation with EU governments at this time.

## Technological Bias

Because the physical operational nature of human smuggling and trafficking are similar (illegal movement across international borders), I surmise that the tools the company uses to track and remove content related to human trafficking are likely the same tools they would use to track and remove content related to smuggling. In a NextGen Safety Tech Talk held at Facebook headquarters in March 2018, experts detailed some of the anti-trafficking interventions the platform employs:

1. Algorithms – detect patterns of interests and participation in illicit activity on the platform.
2. Content Moderators – review and removal of illegal content by trained reviewers.
3. Hashing – attributing unique PhotoDNA to images for future identification and legal use.
4. Cyber Reporting – reporting of images to cyber hotlines where they are stored for future evidence in criminal cases.

The use of these tactics in the context of smuggling deterrence raises a host of ethical concerns with relation to data mining and the privacy of migrant users. If a social media company were to use data mining to track content related to smuggling services and advertisements, they would inevitably identify the profiles of migrants inquiring about these services, looking at these posts, or otherwise interacting closely with this material. Fleur Johns has shown that both supervised and unsupervised data mining processes are susceptible to bias (Johns 2017: 14). Because data mining programs are usually accomplished by programmers with highly specialized skills, rather than by government officials with knowledge in relevant immigration law and policy, data mining processes can be created with little thought to democratic principles. In supervised data mining situations, the personal bias of the designer, or biases in the policies for which the algorithm will operate, may be reflected in the ultimate design. Unsupervised data mining may form natural biases, as processors cluster certain information in ways that are difficult for even the designer to understand (ibid.: 4). One can imagine a scenario in which a data mining algorithm is used with the final aim of taking down all smuggling-related content, and, in the process, the algorithm may identify and remove posts or pages related to asylum, visa applications, and irregular migration, as such posts may be directly or indirectly linked to content related to smuggling. The social media company would then find itself having to determine the legality and legitimacy of all content related to irregular migration, which would dramatically change the sociological architecture of the platform.

While the DSA states that users whose content is removed would have recourse to appeal the decision, the imbalance of power relations between irregular migrants and government authorities would likely deter migrant users from seeking recourse at the risk of exposing their identity or decreasing their chances of gaining legal status in the EU (Committee on Civil Liberties, Justice and Home Affairs 2020: 6). Thus, it will be increasingly important for social media companies who utilize data mining algorithms for content moderation in this area to question the logic of their design as it relates to the autonomy, respect, and privacy of migrant users. It is by no means impossible for algorithms to be designed to detect smuggling content while sifting through content related to irregular migration, but the issue at hand is the potential of these technologies, when designed under intense legal pressure, to profile and possibly infringe on the rights of migrant users.

## VIRTUAL BORDER ENFORCEMENT

Finally, I consider how internet governance relates specifically to the evolution of immigration and border securitization politics by analysing the materiality and purpose of borders, the role of content moderation in the mediation of migration politics, and how the negotiations of responsibility by ISPs and governments may transform social media into coercive infrastructures on which practices of what I refer to as 'virtual border enforcement' are played out.

According to Huub Dijstelbloem, the purpose of borders is to 'mediate between states and between people who are allowed to enter and those whose access is denied' (Dijstelbloem 2016: 311). In critical border studies, borders themselves have come to be understood not just as the physical barriers that demarcate territories, but as 'variegated architecture(s) of control' (ibid.: 314). In other words, borders are the conceptual and material infrastructures of power and control that consist of various technologies and institutions (ibid.: 317). Dijstelbloem contends that borders are extended or reinforced through 'infrastructural events' or 'the moments and . . . specific locations where territory, state authority, jurisdiction, migrants and technologies intersect' (ibid.: 311). Social media platforms are such locations, as states and ISPs clash over jurisdiction and authority over migrants' and smugglers' user-generated content and rights related to the technology at the macro level, and migrants mobilize and facilitate migration at the micro level. Therefore, social media is a site where virtual borders are erected to deter mobility associated with user interaction and user-generated content. I consider below how this virtual border is maintained and enforced.

The DSA represents the transformation of content moderation into a 'technology of power' (Lindskov Jacobsen 2015:123). Technologies of power are technologies which render additional parts of human identity visible, and therefore governable (ibid.). When content moderation is used to monitor, censor, and remove the legal speech of migrants related to irregular migration, the result is not only the removal of knowledge on the migrant experience, but also the demarcation of migrant users' digital and political identities (ibid.). Content moderation is co-opted or enforced by states to produce new 'truths' about migrants' digital identities, experiences, and access to borders (ibid.). The assertion of these truths is an extension of state control over the digital lives of migrants and citizens. The removal and censorship of migration-related content results in the restricting of knowledge about important migratory resources that may ultimately block some migrants from reaching their intended destination or gaining the legal status they seek in the EU. Content moderation thus becomes one of the many technologies that constitute borders. If the goals of deterrence set forth in the EU Action Plan Against Migrant Smuggling are achieved, in part, through content moderation, a case may be made that this represents a state move to externalize EU borders into sociotechnical space.

European governments maintain their externalized borders, in part, by using new technologies as 'coercive infrastructures' (Andersson 2016: 35). These coercive infrastructures are designed to increase cooperation on border enforcement and migration agendas through socio-political relationships mediated through technology (ibid.: 35–6). According to Andersson's research, early smart border technologies on the border of Spain and Northern Africa did not drastically change rates of apprehensions at the border, but the high-tech satellite and radar technology used to police the border required increased cooperation between African states and Spain (ibid.). Thus, border technologies were artifacts through which the socio-politics of migration, deterrence, and border security were mediated. Andersson's conceptualization of coercive infrastructures is also relevant to the politics of internet governance as it relates to migration. Coercive infrastructures are created when two conditions are fulfilled through (1) surveillance or security technologies and (2) the externalization of border patrol responsibilities (ibid.). The proposed DSA would officially legitimize the use of social media as a surveillance technology. It would also result in the mandatory cooperation between social media companies and the EU. The subsequent socio-political relationship between the EU and social media companies is mediated through the technology of surveillance, in this case, the social media platform itself, and the social media company is coerced into doing the work of border enforcement through content moderation. The aim of the legislation, in part, is to externalize border patrol responsibilities to actors with fewer legal restraints towards migrants and their digital rights. Even if Andersson's coercive infrastructures concept does not directly apply to the social media platform, which it may be argued is less an artifact than it is a dynamic public space, the externalization of border politics and policing into social media spaces at the macro level of internet governance certainly has 'coercive outcomes' (ibid.: 35).

I return to the Facebook example to bring these concepts together. If Facebook is not yet using the above tactics to monitor human smuggling content on its platform, it is likely because the company is not legally bound to do so under the E-Commerce Directive and because doing so would raise serious ethical concerns about the right to free speech on its platform. If Facebook did engage in a more active pursuit of smuggling-related content, it could find itself in the politically and morally ambiguous situation of taking down both smuggling advertisements and the migrant community groups described above which enable migrants to navigate their journeys without a smuggler, to access important health and safety resources, and to connect to networks that they rely on during migration and integration into their host community. A company like Facebook would thus find themselves in exactly the situation which has morally plagued the field of smuggling deterrence since its inception; increasing smuggling deterrence measures inevitably limits migrant choice and undermines their agency. Blanket removals of content related to irregular migration would take away an important (albeit limited) form of control that migrants have over certain factors of their journey. It would also silence their narratives about the migrant experiences and their digital voices in sociotechnical space. The social media company, in this case, would be coerced into using their technologies to further state interests in deterring not only smugglers, but all mixed migration towards their borders. This would represent a paradigmatic change in the process of border enforcement and state territorial expansion into sociotechnical spaces; non-state actors would be legally and normatively obligated to act in a way that ultimately bolsters the EU border security agenda. Virtual borders and border enforcement would thus be co-constructed by state law and non-state actors, and maintained through content moderation technologies, effectively externalizing EU borders into public sociotechnical spaces.

## CONCLUSION

In conclusion, in light of high rates of migrant death in attempted border crossing worldwide, it is imperative that we turn our attention towards the factors that bolster or limit migrant safety in transit. The relationship between migrants and smugglers is an obvious start, and while this relationship is a source of significant risk for migrants, it is also a relationship of complexity and, for many migrants, smugglers are a necessary means to a highly desirable end. It is important to remember that demand for smugglers is created by the strict border security regimes which limit the movement and agency of migrants worldwide. More durable solutions to this crisis would call for the drastic restructuring of border and migration policies, such as even burden sharing of asylum seekers among member states and an increased commitment to resettle refugees (which would theoretically obviate or at least alleviate the need to pour so many resources into asylum deterrence) (Gibney 2004: 243). However, as such a reconstruction is unlikely to occur in the near future, and as not all mixed migrants are refugees, it is important to turn our attention towards the ways that mixed migrants use the tools at their disposal to empower and protect themselves when they make these perilous journeys to EU borders.

This chapter has outlined the ways that migrants demonstrate agency through micro-level governance online. Their contributions to the spreading of narratives and information through their individual activity and their interactions with other users create norms about the migrant experience that have the potential to inform the decisions and safety precautions of other migrants. Their use of social media as the conduit for this information is also a political declaration of their right to mobilize and advocate for themselves in sociotechnical spaces. However, while social media holds great potential for migrant agency, it also involves risks from actors besides some disingenuous smugglers. State attempts to deter human smuggling online reveal contentious struggles over internet governance of digital rights and data privacy at the macro level that raise serious concerns about the violation of migrant users' rights to free speech and access to information online and which limit their political, digital identities. It is important to think critically about the changing relationships of governments to ISPs and how this may shape the digital realities that inform, reflect, and challenge our lived realities. For migrants using digital platforms to help navigate their journeys, the stakes involved with the passing of the DSA or similar legislation are significantly high. Thinking of social media as a coercive infrastructure around which border policy and agendas are promoted, and content moderation as a technology of power through which border enforcement is achieved, helps us to understand the gravity of these stakes. Using the concept of virtual border enforcement as a lens to examine this phenomenon also renders visible the dynamic intersections between the political agendas and actions of migrants, smugglers, states, and non-state actors, who all contribute at different levels to the construction of sociotechnical space and to the evolution of immigration politics on and offline.

# REFERENCES

## Academic Works and Reports

Alencar, A., Kondova, K., and Wibbens, R. (2018), 'The smartphone as a lifeline: an exploration of refugees' use of mobile communication technologies during their flight', *Media Culture and Society*, 41(6): 828–44.

Andersson, R. (2016), 'Hardwiring the frontier? The politics of security technology in Europe's "fight against illegal migration"', *Security Dialog*, 47(1): 22–39.

Brooks, B. (2015), *Effects of Organization-level Internet Governance: A Mixed-methods Case Study Approach to Social Media Governance*, ProQuest LLC.

Castro, D. and Chivot, E. (2019), 'What the EU should put in the Digital Services Act', Center for Data Innovation.

Dijstelbloem, H. (2016), 'Borders and the contagious nature of mediation'. In K. Smets, K. Leurs, M. Georgiou, S. Witteborn, and R. Gajjala (eds), *Sage Handbook of Media and Migration*. London: Sage, pp. 311–20.

Frouws, B. and Brenner, Y. (2019), 'Hype or hope? Evidence on use of smartphones and social media in mixed migration', Mixed Migration Center.

Frouws, B., Phillips, M., Hassan, A., and Twigt, M. (2016), 'Getting to Europe the "WhatsApp" way: the use of ICT in contemporary mixed migration flows to Europe', Danish Refugee Council.

Gibney, M. (2004), *The Ethics and Politics of Asylum: Liberal Democracy and the Response to Refugees*. Cambridge: Cambridge University Press.

Global Alliance Against the Trafficking of Women (2011), 'Smuggling and trafficking: rights and intersections', GAATW Working Paper Series 2011.

Greenwood, F. et al. (2017), 'The Signal Code: a human rights approach to information during crisis', Harvard Humanitarian Initiative.

Hoffman, A. and Gasparotti, A. (2020), 'Liability for illegal content online: weaknesses of the EU legal framework and possible plans of the EU Commission to address them in a "Digital Services Act"', Centre for European Policy.

Johns, F. (2017), 'Data mining as global governance'. In R. Brownsword, E. Scotford, and K. Yeung (eds), *Oxford Handbook of Law, Regulation and Technology*. Oxford: Oxford University Press, pp. 776–94.

Kolodziejczyk, M. (2015), 'Liability of internet users and internet data providers in the context of the reform of the EU Data Protection Law', *ASEJ Scientific Journal Bielsko-Biala School of Finance and Law*, 4(6): 57–79.

Latonero, M., Poole. D, and Berens, J. (2018), 'Refugee connectivity: a survey of mobile phones, mental health, and privacy at a Syrian refugee camp in Greece', Harvard Humanitarian Initiative, Data & Society Series.

Lindskov Jacobsen, K. (2015), *The Politics of Humanitarian Technology: Good Intentions, Unintended Consequences, and Insecurity*. Routledge Studies in Conflict, Security and Technology. London: Routledge.

Mixed Migration Center (2018), 'Experiences of female refugees & migrants in origin, transit, and destination countries: a comparative study of women on the move from Afghanistan, East and West Africa', Mixed Migration Center.

Roberts, Z. (2017), Information exchange between smugglers and migrants: an analysis of online interactions in Facebook groups. Criminal Justice, Borders and Citizenship Research Paper No. 3051186.

Sanchez, G. (2017), 'Critical perspectives on clandestine migration facilitation: an overview of migrant smuggling research', *Journal on Migration and Human Security*, 5(1): 9–27.

## Legal and Policy Documents

Committee on Civil Liberties, Justice and Home Affairs (2020), Draft Report on the Digital Services Act and fundamental rights issues posed (2020/2022(INI)). Presented to European Parliament, May 2020.

European Commission (2015), Communication from the Commission to the European Parliament, the Council, and the European Economic and Social Committee of the Regions of the EU: Action Plan against migrant smuggling (2015–2020).

European Commission (2016a), EMN Inform: the use of social media in the fight against migrant smuggling. 2015–2020.

European Commission (2016b), EMN Ad-Hoc Query on COM AHQ on Addressing and preventing the use of social media in migrant smuggling – exploring cooperation frameworks with social media and other relevant online service providers. Requested by COM on 18th April 2016 Irregular Migration. European Migration Network.

European Parliament and the Council of the European Union (2000), *Directive 2000/31/EC of the European Parliament and of the Council on June 8, 2000 on certain legal aspects of information society services, in particular electronic commerce, in the Internal Market* ('Directive on electronic commerce'). Official Journal L 178, 17/07/2000. Luxembourg.

European Union (2016), Directive (EU) 2016/680 of the European Parliament and of the Council of 27 April 2016 on the protection of natural persons with regard to the processing of personal data by competent authorities for the purposes of the prevention, investigation, detection or prosecution of criminal offences or the execution of criminal penalties, and on the free movement of such data, and repealing Council Framework Decision 2008/977/JHA.

UN General Assembly (2000), *Protocol against the Smuggling of Migrants by Land, Sea and Air, Supplementing the United Nations Convention against Transnational Organized Crime*, entered into force in Geneva on 15 November 2000.

## News Articles

Adamson, D. and Akbiek, M. (2015), 'The Facebook smugglers selling the dream of Europe', BBC News.

Greenfield, C. (2020), 'As governments build advanced surveillance systems to push borders out, will travel and migration become unequal for some groups?', Migration Policy Institute.

Human Rights Watch (2019), 'Rights group warns against DHS's use of social media monitoring of immigrants', Human Rights Watch.
RT News (2016), 'Migrant smugglers use Facebook to promote Turkey-Italy trips bypassing sealed Balkan route', RT News.

## Other

Article 10. Human Exploitation. Facebook Community Standards. Accessed March 2020.
NextGen Safety Tech Talk. Global Safety Summit. Held at Facebook headquarters, March 2018.

# 6. Irregular mobility and network capital: the case of the Afghanistan-Iran smuggling route

**Ruta Nimkar, Emily Savage and Abdullah Mohammadi**

## INTRODUCTION

Social networks of migrants play a central role in intentions to migrate, route choice, smuggling method, and destination country (Haug, 2008; Dekker and Engbersen, 2014). This plays out in a circular manner: successful migrants facilitate prospective migrants, encouraging further prospective migrants, and reducing risks and costs associated with movement (Broeders and Engbersen, 2007). To date, literature has focused on formal migration in the context of elites who have access to significant amounts of capital (Bergman et al., 2009, 35), with a greater focus on irregular migration in the last decade with the increase in irregular flows into Europe from North Africa, the Middle East, and Central Asia.[1]

Irregular migrants are increasingly making use of mobile technology to strengthen social networks and support their journeys, transforming the migration process for migrants (Dekker and Engbergsen, 2012; Crawley et al., 2016a; Frouws et al., 2016; Gillespie et al., 2016; Zijlstra and van Liempt, 2017; Brenner and Frouws, 2019). This is on-trend with improvements in accessibility and functionality of social networking and communication applications (e.g., WhatsApp, Viber, Twitter, Facebook). To this end, lack of appropriate technology can now be seen as a barrier for migrants (Sanchez, 2017) – particularly those lacking smartphones (Brenner and Frouws, 2019).

The way in which mobile technology contributes to migrant decisions appears to revolve around strengthening linkages within existing networks (notably family, friends, and smugglers from the same ethnic community) rather than provision of new information. Several reports from the Danish Refugee Council indicate that the most important source of information for people on the move are friends, family, and smuggling networks (Brenner and Frouws, 2019). Some studies suggest that as few as 6 per cent of migrants use social media as a dominant source of information to prepare for the journey (Borkert et al., 2018); however, it functions as a way of maintaining communication with those who provide information and services. That is, social media provides the network capital – bridging capital in particular – to allow people to move. The bridging effect of social media is strong enough that some studies indicate there is a possibility for social media to support an increase in the number of unfacilitated journeys – thereby bypassing smugglers (Gillespie et al., 2018).[2]

Yet, the majority of irregular migration still takes place through smuggling networks (UNODC, 2018).[3] Smuggling businesses take multiple forms, from hierarchical groups with centralised organisational structures to networks of loosely organised groups and individuals (UNODC, 2018). Those within smuggling networks are often closely linked to those who purchase smuggling services; there are often shared ethnic, clan, or familial links between migrant and smuggler (UNODC, 2018). There is an acknowledgement that social networks both between migrants and smugglers and between smuggling networks are a critical facilitating factor in movement and that these networks appear to make increasing use of mobile technology to organise their activities (Sanchez, 2017; Gillespie et al., 2018; Milivojevic, 2018).[4] The existing research on smuggling and technology focuses on the relationship between migrants and smugglers and covers two primary uses: service advertising and communication (Musto and Boyd, 2014; Hacsek and Visnansky, 2017; UNODC, 2018). However, there is less research, both in the academic sphere and in the policy and practice realms, regarding the ways in which information technology affects these networks. The lack of research into this topic reflects a more general dearth of information on smuggling networks more broadly, driven by practical access constraints (IOM, 2018; UNODC, 2018).

## Framing Technology, Social Media, and Smuggling

Network theory demonstrates that the ways in which individuals interact and network within their social system helps to develop payoffs in terms of social capital (Lin, 1999) through increasing the strength and number of relationships and acquaintances (Bourdieu and Wacquant, 1992). Network theory contends that migration is linked to personal capacities and characteristics at the individual migrant level but also to the relationality of individuals (Bergman et al., 2009; Urry, 2012). The theory was developed in a context characterised by technological developments in the field of transport and communications (Koslowski, 2011), including the uptake of mobile phones and social media platforms. Network theory provides a framework for understanding how technological developments affect the ways in which individuals relate to their communities and, as such, provides an appropriate framework for analysing migration patterns. It 'acknowledge[s] that moves tend to cluster, can be circular, and take shape within wider contexts and systems' (O'Reilly, 2015, p. 5). Mobile phones and social media play a catalytic role in mobility, increasing access to networks and facilitating network ties (Rettie, 2008; Grabowicz et al., 2014).

Urry (2007) outlined eight enabling factors that help to support relations between individuals and their networks and to facilitate mobility, wherein people experience stratified 'mobility regimes' on the basis of the varying access they have to the eight elements of network capital (Gillespie et al., 2018). This in turn determines their access to specific resources. Smuggling networks offer migrants access to a parallel and irregular mobility regime. Smugglers are known to offer services such as access to physical and informational movement capacities, access to safe and appropriate meeting places, and access to transportation (UNODC, 2018). As such, they provide access to network capital for those who may not have such access through formal means. The elements of Urry's (2007) conceptualisation broadly apply to regular and irregular migration routes, with some adjustments. For example, Urry's (2007) Element 1 is predicated on access to, and privilege in, the formal migration system; smugglers offer, instead, a network which enables this system to be bypassed through other social networks, bribes, and other means of facilitation that allow the passage through international borders. This chapter uses an adapted version of this model to analyse the effects of technology on smuggling networks (Table 6.1).

The chapter analyses the effects of mobile technology on smuggling networks using the basic assumption that smuggling organisations are developing (intentionally or not) an alternative mobility regime for those who cannot access the standard mobility regime. The regular mobility regime can be described through the framework of Urry's (2007) eight elements of network capital; this framework can also be adapted to irreg-

*Table 6.1* *Eight elements of network capital (Urry, 2007) adjusted to migrant smuggling*

| Network Capital Element | Smuggling Adjusted Elements |
|---|---|
| An array of appropriate documents, visas, money, qualifications that enable safe movement | An array of networks within and information about the relevant authorities (official and unofficial) to permit safe movement |
| Social contacts at a distance offering hospitality and invitations | Social contacts offering connections that are relevant to the next phase of the route |
| Physical and informational movement capacities | Physical and informational movement capacities with minimal oversight from authorities |
| Location-free information and contact points such as real or electronic diaries, answer phones, mobile phones, email, etc. | Location-free information and contact points with minimal oversight from authorities |
| Communication devices | Unchanged |
| Appropriate and safe meeting places | Unchanged |
| Access to means of transportation | Unchanged |
| Time and other resources to manage and coordinate the other seven elements | Unchanged |

ular migration routes. For the purposes of this chapter, the elements are then clustered into three groups, reflecting components of a smuggling network:

- *Relationships between smugglers and migrants*. Generating new business and delivering the smuggling service to clients depends on the capacity of smuggling networks to engage with potential migrants in order to disseminate information and marketing material to potential migrants and to attract the migrants to the services they offer. This is linked to Element 3.
- *Relationships within smuggling networks*. In order to sustain themselves, smugglers need to develop an informal network consisting of, at minimum: information and contact points, communication devices, safe and secure meeting places, transportation, and time. This reflects Elements 4, 6, 7, and 8.
- *Relationships between smuggling networks and external actors (other networks and authorities)*. Elements 1 and 2 reflect engagement with external networks including authorities such as the border police and armed forces, as well as other smuggling networks.

As the 'regular' mobility regime is described through the framework of Urry's eight elements of network capital, it can also be adapted to irregular migration routes to understand supply, organisation, and demand.

The chapter focuses on small-scale smuggling, defined here as decentralised, local-level smuggling networks that rely primarily or exclusively on family and ethnic ties and provide basic services (guiding, driving, accommodation) to migrants. The chapter adopts the movement of Afghans across the Afghanistan-Iran border towards Europe as a case study.

## METHODOLOGY

This study builds on a previous study conducted by the same authors (Mohammadi et al., 2019) by engaging an additional eight Afghan smugglers – seven in Kabul and one in Herat – ranging in age from 26 to 53 years old. Of the smugglers identified, the researchers made an effort to select those from diverse ethnic background, ages, monthly client figures, and relative extended network sizes. Smugglers identified represent different segments of the smuggling economy in terms of services provided and the 'level' of service (e.g., visa purchases and direct flights to overland travel through multiple connected networks). The smugglers were identified through migrants and sources known to the authors as well as the researchers' networks with international non-governmental organisations working on mixed migration data collection.

The data collection was made possible because the researchers have significant experience in and engagement with communities affected by migration. The researcher responsible for data collection has been collecting migration data in Afghanistan since 2016. The research team has close links with community members within these areas, including with local officials, migrants, logistics service providers, and moneylenders. The team drew on these connections when identifying potential interviewees.

The researchers collected data using semi-structured interview formats and analysed data using inductive coding methods. Data collection teams comprised two people, the lead researcher and a field assistant, both male. Interviews were generally held at the interviewee's home, office, or workplace and were held in the local language (Dari). Interviews took place in January 2019. Interviews were conducted with full informed consent, but most interviewees did not permit the researchers to record the interviews or to refer to their names in the research.

The research has several limitations. First, as much irregular migration and smuggling research, the illegal aspects of the subject matter limit the sampling strategies available and reduce the number of informants available to researchers. Thus, a snowball sampling strategy that relied on gaining trust from individuals to share sensitive information was employed. Due partially to travel schedules, fewer informants were identified in the border areas than anticipated. Furthermore, this research is primarily engaged with smugglers who operate along the Western smuggling route via Iran and Turkey and onwards to Europe, rather than the Eastern route through Pakistan – so it is possible that the research does not fully represent a picture of mobile technology and irregular movement in Afghanistan as a whole. Despite the limitations, the perspectives of smugglers are a significant gap in the literature and thus this research offers limited but important insight into smuggler engagement with social media.

## SOCIAL MEDIA AND NETWORK CAPITAL BETWEEN SMUGGLERS AND MIGRANTS

The relationship-building process between smugglers and migrants consists of two phases. In the first phase, smugglers and migrants gain information and make first contact with one another. This phase can include direct advertisement where smugglers disseminate information on the services they provide, and migrants initiate contact based on the information provided or direct contact is made through word-of-mouth referrals. The second phase consists of more concrete interactions in preparation for movement, during movement, and upon arrival at the destination or transit point where the migrant will be transferred to another smuggling network for onward travel.

Increasingly, mobile technology allows for a feedback loop between the two phases where migrants who have good experiences provide feedback to potential migrants as do those who have bad experiences (Tinti and Reitano, 2017). Previous analysis regarding the relationships between migrants and smugglers has focused on the migrant's point of view (e.g., Crawley et al., 2016b; Dekker et al., 2018; Gillespie et al., 2018). The following section details the changing nature and mechanisms of relationships between migrants and smugglers from the point of view of smugglers. By and large, this validates existing migrant-centric research in that mobile technology and social media have enhanced the network capital element of physical and informational capacities and can do so without oversight from authorities.

## Social Media and Advertisement of Smuggling Services

Social media platforms have been noted to spread unrealistic and false migration experiences to encourage potential migrants to pursue movement (Dekker and Engbersen, 2012; EMSC, 2018; Gillespie et al., 2018). A bulk of the research focuses on diaspora populations using social media and creating a pull factor for other potential migrants (Frouws et al., 2016; RVI, 2016). Studies that examine advertisement and recruitment directly from smugglers often focus on the content and tactics of advertisements rather than how relationships are built (Roberts, 2017). Advertisement takes place primarily over 'public' social media such as Facebook and Instagram (Sanchez et al., 2018), though may be quite subtle. Sanchez et al. (2018) point out that most Facebook posts consist of carefully curated self-representations of migrants and smugglers in Europe. Smugglers have also been found to advertise 'trip packages' online through social media and messenger groups (EMSC, 2015).

Interviews conducted for this study generally reinforce the findings of existing research. Most smugglers mention using wider social media outlets such as Facebook and Instagram for reaching a large audience and then redirect individual conversations to messenger services such as WhatsApp and Viber.

> Facebook and Telegram are very useful in terms of advertisement … There is no need to be straight about your business mentioning that you cross people illegally. Just mention migration opportunities and people will know what you mean … in Telegram we have made some groups for consultation about migration, education and health visas, giving information about migrants in Germany, Sweden, Norway, etc. On Facebook also we have several pages that do the same. (Male, 26, Hazara, Interviewed in Kabul)

Seven of the eight smugglers who responded to questions about whether there was false advertising in the social media market agreed that false advertising was a problem, though perspectives on migrants' willingness to be deceived varied.

> Some other networks are trying to find clients through Facebook and Telegram channels, and they say what they have to say to find new migrants … yes, they tell lies, they say that there is work in Turkey, but we know that

> there isn't such a thing. Many of migrants are wandering in Turkey without job and a roof to sleep under. (Male, 40, Tajik, Interviewed in Kabul)

While some of the smugglers interviewed indicated that migrants come to smugglers with unrealistic expectations based on advertisements, others felt the bulk of advertising came from migrant friends and family members posting positive stories and photos of their journeys – not as a result of images propagated by individual smugglers or smuggling networks themselves. Others conceded that smugglers have no choice but to promote positive images and stories to attract clients.

> You have to do that to attract people's attention. Otherwise they just skip the ads and go to someone else. Of course, it's not false information – we just encourage them to migrate by posting beautiful pictures of European countries. Europe is beautiful and with full security. So, it's technically not a lie or false information. I personally think that the youth should migrate from Afghanistan at any cost because their lives are wasted here. At least there, they could have a normal life far from explosions, suicide attacks, or unemployment. (Male, 28, Tajik, Interviewed in Kabul)

It is widely understood that when smugglers are advertising, the alternative to social media platforms is ethnic, clan, and familial networks (Monsutti, 2008; Frouws et al., 2016). Many smugglers still rely heavily on these networks to conduct advertising and find clients (Stone Cadena and Velasco, 2018), with some smugglers using personal recruitment methods (RVI, 2016). Nonetheless, in areas where mobile technology is being taken up, social media is providing a new outlet for advertisement and gaining clients that otherwise would have been out of reach of many low-level smugglers.

Advertisement of smuggling services to a market of actual and potential migrants reflects an improvement in information available to migrants. The use of digital technology and social media platforms helps to spread the reach of advertising messages; as such, it improves informational movement capacities, that is, Element 3 of social capital.

### Messenger Apps and Organisation of Journeys

Mobile technology also facilitates more direct communication between different groups on irregular migration paths. Private platforms such as Telegram and WhatsApp are frequently used to organise logistics (Haczek and Visnansky, 2017; Brenner and Frouws, 2019). Prior to low-cost mobile technology, smugglers and migrants had short-term interaction, whereby the migrant and smuggler were not likely to be in contact after the migrant passed to a new geographic area. Migrants have always heavily relied on information provided by smugglers (Koser and Pinkerton, 2002; Zijlstra and van Liempt, 2017), and technological advancements and social media have facilitated longer-term interaction to verify information, understand next steps, share contacts, and make payments (Frouws et al., 2016).

Furthermore, social media – and improved access to mobile technology more broadly – has begun to shift the balance of power; migrants can now check on the quality of smuggling services from a wider source of information, including reviews and recommendations by other migrants through direct links made with successful migrants over social media or through public channels such as Instagram and Facebook. Though not all of the information is high quality, this may support a feeling of empowerment among migrants. Some studies suggest that migrants may even monitor smugglers in real time against planned routes (Gillespie et al., 2016). Through these increased checks, studies suggest that migrants may place an increased level of trust in their smugglers as smugglers have an interest in maintaining 'positive reviews' and attracting more clients (Triandafyllidou and Maroukis, 2012; Roberts, 2017; Zijlstra and van Liempt, 2017). Some interviewees in this study confirmed this conclusion, suggesting that, despite having ultimate control over the information, it was not in their interest to misrepresent the realities of the journey to clients due to the changing communication environment.

> Many smugglers give false information about the situation on the route or in the destination countries, but we don't. It's bad for the business, because most of our customers are introducing us to other migrants especially to people in their community back in Afghanistan. So, if we give false information, we will lose our future customers. (Male, 33, Hazara, Interviewed in Kabul)

> Look, every route has its own difficulties. If we don't tell the migrants about them, we ourselves will face problems. We tell them that they have to walk for 16 hours or they have to get warm clothes … We try to keep them as much as possible informed about what is ahead of us so they are prepared and not get surprised. Their problem is our problem. (Male, 28, Herati, Interview in Herat)

The practice of providing detailed information on journeys is consistent with other studies in the region (Koser, 1997; van Liempt, 2007) and suggests a fairly consistent practice of providing specific information that the smuggler alone would have access to. Furthermore, past research has suggested that the personal or community relationships between smugglers and smuggled is crucial in determining the level of information and the accuracy of information provided to migrants. It has consistently been shown that when migrants and smugglers share similar social networks, the migrant is less likely to be deceived (van Liempt, 2007; Monsutti, 2008). Likewise, exploitation may be more likely in cases with low network ties and where business interests supersede altruistic motivations (Dekker and Engbersen, 2012). In the case of Afghan smugglers explored here, linkages and relationships that exist electronically do have the potential to serve similar roles as traditional local networks.

While migrants certainly face very real protection risks or even death and thus receiving forms of assurances and trust building is paramount to decision-making behaviour, smugglers also face risk of arrest, detention, and extortion as they are participating in an illicit economy. Smugglers interviewed for this research highlighted this challenge – suggesting social media offered them the same assurances and risks as potential migrants. Specifically, social media has allowed smugglers to more easily trust potential clients. Smugglers are concerned with the background of their clients to ensure their legitimacy and reduce their exposure to sting operations. Smugglers report being able to check backgrounds of clients (presumably using Facebook or other networking applications) and extensive conversations could take place before ever meeting in person. This included sensitive conversations about costs and arranging payment.

Other smugglers suggested that, although state surveillance through accessing conversations on mobile applications was less of a risk than previous methods, impersonation by authorities or criminals was something that ultimately presented the same risks as in the past. Smugglers also highlighted that, although social media provides a different set of assurances to a larger clientele base, it has not necessarily increased trust to a level comparable with traditional direct or communal ties or to a level that would ultimately change the power imbalance between migrants and smugglers completely:

> Trust is when you know someone … the migrants don't know me – it doesn't matter if he sees me in person or on Facebook or Telegram. (Male, 26, Hazara, Interviewed in Kabul)

> The migrants are afraid of us because their money and their lives are in our hands. They have to do whatever my colleagues tell them otherwise they will

> be in trouble. So, what is here: it's not trust – it's fear. (Male, 53, Pashtun, Interviewed in Kabul)

The capacity to discuss smuggling routes, prices, and mechanisms in a safe and secure fashion represents an improvement in availability of information, as well as an improvement in the type of information provided. The possibility for discussion ensures that migrants have access to a greater amount of more tailored information from smuggling networks. As such, messenger applications support informational movement capacities, and, through this, help to support the alternative mobility regime provided by smuggling networks.

## SOCIAL MEDIA AND THE INTERNAL DYNAMICS OF SMUGGLING NETWORKS

An individual smuggling organisation may provide an array of services, including planning, transportation, guiding irregular border crossing, accommodation, networks and information to evade capture by local authorities, and fraudulent travel documents (UNODC, 2018). In order to provide these services, smuggling networks require internal network capital. Internal network capital allows smugglers from within the same network to engage and coordinate between themselves in order to provide services.

Key elements of the infrastructure required to organise smuggling networks include information and contact points, communication devices, meeting places, and accommodation (UNODC, 2018). Smuggling networks tend to consist of groups of friends and family members, often linked by ethnic or clan ties (Sanchez, 2017; UNODC, 2018), and often consisting of individuals who have themselves been (irregular) migrants (Sanchez, 2017; Stone-Cadena and Velasco, 2018). Current research suggests that smugglers rely on mobile phones for their operation (McAuliffe in IOM, 2016; Newelle et al., 2016).

In the primary data collection for this chapter, the researchers considered the ways in which mobile technology and social media changed internal dynamics within an individual smuggling network and have been found to have significantly improved internal organisational network capital for smuggling organisations. According to the modified network capital framework, improved location-free information and contact points have been made available, and improved communication and transportation coordination can now be organised – ultimately freeing additional time and capacity to move more migrants. These changes come about in light of reduced cost and improved speed of operations facilitated by cheap and accessible phones and secure social media applications. Eight of the ten smugglers pointed out that technology made smuggling operations cheaper, due to decreased communication costs. One smuggler added that improved communication reduced the necessity for smugglers to travel, thus making the operation more profitable. Similarly, six of the ten smugglers discussed the fact that mobile technology made operations more convenient and faster.

> It's not comparable with five year ago at all. Five years ago, it was only phones and satellite phones that we can communicate with each other. It was expensive, especially satellite phones, and also very dangerous. (Male, 45, Pashtun, Interviewed in Kabul)

> Many places didn't have phone coverage, so we had to use a Thuraya phone [satellite phone] which is much more expensive than ordinary phones. With Whatsapp and Viber, it's easy and everywhere is covered by the network. (Male, 40, Tajik, Interviewed in Kabul)

Coordination of services within a network becomes easier to organise and less costly as members of networks can now be based in diffuse areas and easily gain information about movements from other partners. Smugglers have noted that it is less necessary to send a person to accompany a caravan or to organise specific elements of a trip – these aspects of the trip can now be organised and monitored remotely. This reduces the costly lack of coordination between facilitators that has been noted in other studies (McAuliffe in IOM, 2016; Sanchez et al., 2018). Improvements in communication affect a wide variety of services across the spectrum in which smugglers operate. Interviewees included travel agents, money transfer agents and facilitators; eight out of ten interviewees indicated that mobile technology had improved communications within their own networks. Respondents came from a variety of backgrounds, and also referred to several different services in their responses.

> Before, arrangement of the journeys was very difficult. For example, to provide accommodation along the way, you have to call and one of our colleagues must go there before the caravan reaches there to prepare the place, buy or prepare the food if necessary and do other tasks. But now, it's only calls. (Male, 33, Hazara, Interview in Kabul)

Mobile technology and social media have also improved the division of tasks within a network. Past studies indicate that smuggling facilitators often work independently, often performing highly specific tasks (Campana, 2016; Zhang et al., 2007). Facilitators often do not know each other, and the cost of meeting is also significant (Sanchez, 2017). Some smugglers take on coordinating roles in which they monitor journeys, put people in contact, support logistics, and regulate financial transfers. For the smugglers who take on coordinating roles, mobile technology provides a mechanism to improve communication. This has already been pointed out by some academics (Sanchez, 2017) and is strengthened by the interviews conducted for this piece of research.

> It's not like before that everything was arranged by phone and we all accompany the caravans to destinations. Now, we can easily distribute the tasks among ourselves and make the operations more efficient and safe – as much as possible. (Male, 53, Pashtun, Interviewed in Kabul)

Mobile technology has also improved the logistics systems needed to operate an organisation, including improved money transfer practices. Money transfer often takes place through informal money transfer mechanisms (*hawala*) that rely on familial and ethnic linkages for their trustworthiness. Increased usage of mobile phones has resulted in same-day money transfers through the informal *hawala* system, which, prior to the prevalence of smartphones, would have taken around one week (Male, 33, Hazara, Interviewed in Kabul). By improving communication for money transfer networks, mobile technology has improved the overall infrastructure for smuggling networks, facilitating business and attracting prospective clients.

## SOCIAL MEDIA AND NETWORK CAPITAL BETWEEN SMUGGLERS, SMUGGLING NETWORKS, AND AUTHORITIES

To provide adequate service, migrant smugglers and networks engage with two primary external networks: other smuggling networks and state authorities. Engagement with other smuggling networks allows smugglers to provide a smooth service in which migrants transition relatively seamlessly from one network to another to proceed as far as possible along the smuggling route. Smuggling networks may also engage authorities or monitor them with the intention of complete avoidance to reduce the risks associated with irregular crossing (e.g., physical assault by border guards or detention) and ensure successful crossing (UNODC, 2018). Technology and social media have significantly improved the capacity of networks to share information and expand social contacts to pass migrants onwards along the route with limited or no cost, across international boundaries, and without oversight from authorities – ultimately reinforcing Elements 1 and 2 of the modified network capital framework.

However, the effect of technology and social media on the relationship between different smuggling networks is significantly different from the effect of technology and social media on the relationship between smuggling networks and authorities. While mobile technology appears to have significantly improved operational links between smuggling networks, it has had no significant effect on relationships with authorities. Technology and social media have instead presented unique benefits for evasion of authorities but also new risks for capture and punishment.

## Other Smuggling Networks

In order to move migrants across long distances, smugglers must form associations and links with different networks, often with different family and ethnic origins (UNODC, 2018). Trust between different smuggling networks is assumed to be low (IOM, 2016, 2018), and communication and engagement can be challenging due to geographical circumstances. There is a general lack of data regarding how different smuggling networks communicate between each other, linked to the overall challenges associated with gathering data on smuggling (IOM, 2016).

The smugglers interviewed in this research were clear that mobile technology and social media applications have had distinct benefits for interactions between smuggling networks. Out of ten respondents, only one said that relationships with other smuggling networks were unchanged. Most respondents stated that mobile technology improved communication around logistical arrangements and improved clarity in relationships and roles between smugglers but only half indicated that mobile technology increased trust between smuggling networks. Those who argued that technology increased trust cited multiple reasons including improved knowledge about the state of the route and the security conditions. Similar to providing a check on the migrant-smuggler power imbalance, some smugglers indicated that improved technology opened up a new punishment mechanism for those who were not trustworthy as poor reviews could be spread quickly. Those who believed that mobile technology had no effect on trust between smuggling networks based their opinions primarily on the fact that exchanges between networks remain transactional in nature.

> [Mobile technology] made the communication between different networks easier but hasn't increased the trust because it's a business and when it comes to business and money you cannot trust your brother. (Pashtun, 45, Kabul)

Two of the respondents who indicated that trust would remain constant indicated that trust is formed through longer-term links and personal interactions. One of these respondents indicated that ethnic and family links are predominant in establishing trust. In this case, mobile technology can smooth day-to-day interactions, and overcome the barriers of ethnicity and nationality.

## Authorities

Migrant smuggling involves engagement with local authorities and, generally, a degree of complicity. One study on smuggling routes from Somalia to Yemen found that government authorities owned some of the ships used for smuggling (REF, 2017); data from the Danish Refugee Council on relationships between smugglers and authorities in Central Asia and the Horn of Africa found that one of the most prevalent forms of protection violations was bribery, directed primarily at local authorities. The relationship between smuggling networks and authorities has not been explicitly analysed; this again is due to the fact that there is limited data about smuggling (IOM, 2018).

The smugglers interviewed for this research indicated that contact with authorities often takes place face-to-face and that this is unlikely to change in the near future. The reliance on face-to-face communications with authorities who accept bribes is likely due to the fact that authorities are aware that communication may be traced. One smuggler detailed the ways in which visas can be purchased from officials by smugglers and, given the details of these interactions are complex, are easier to negotiate in person.

Smugglers universally highlighted that social media made their work more secure by providing a free means of encrypted communication that could easily be deleted as often as needed to maintain a 'clean' device. Furthermore, informants mentioned that increased communication amongst networks in general had helped smugglers understand in real time changes in roads and border crossing conditions, thereby reducing interactions with authorities.

> … the risk of attracting attention of security officials or getting caught by police is reduced. The government have no control over these media. While when it comes to phone calls, they can control it. This is true mostly for our colleagues in Iran. (Male, 40, Tajik, Interviewed in Kabul)

> If it [the app] gets dangerous because of government control, we change it to another secure app. For example, after the Iran Government filtered Telegram, we start using IMO there because it's safer. (Male, 28, Herati, Interview in Herat)

Most respondents also suggested that mobile phones presented a new risk. Five of those interviewed pointed to security and authorities as the most significant risk associated with adopting new technology. Smugglers are aware that authorities often inspect phones and with the smuggling business being conducted primarily through applications, detailed information about clients, plans, and networks can be uncovered from a single device.

> Last year, we made a group in WhatsApp among our network to share our information during journey and coordinate it. But once one of the caravans were caught and arrested and the phone of our colleague was seized by the police and it made a lot of trouble for us, because our group chats were all there. So, from then we only chat one-by-one. Because of the security of ourselves. (Male, 40, Tajik, Interviewed in Kabul)

Despite the universal recognition of the risks, very few smugglers suggested grave concern or put in place mitigation measures beyond deleting messages from the applications, with only one taking regular precautions to erase his digital record through replacing his phone every two months. Many seemed resigned to the risks inherent in smuggling: '[I] Just trust in god that such thing won't happen' (Male, 45, Pashtun, Interviewed in Kabul). One smuggler pointed to bribery as a risk mitigation mechanism, remarking that 'many times people from national security office came here and asked about our activities and every time we gave some bribes to not put ourselves in trouble' (Male, 28, Tajik, Interviewed in Kabul). This same smuggler recognised that this mitigation mechanism was unsustainable, adding, 'but until when? They will come back and harass us. It's possible that they will come tomorrow and get all our documents and laptops.'

## CONCLUSION

This chapter makes the general hypothesis that smuggling organisations create a parallel mobility regime and that network capital makes this parallel regime functional across core relationships necessary to operate smuggling networks. On the basis of this hypothesis, it has adopted a network capital approach to analyse the ways in which mobile technology have altered the operation of smuggling networks and suggested a modification to Urry's (2007) eight elements of network capital. The modified elements can support improved understanding of how technology can affect smuggling networks more broadly.

The chapter concludes that mobile technology has reduced cost and increased speed and effectiveness of smuggling organisations. Reduction in cost arises from improved capacity to communicate and coordinate within and between smuggling networks, as well as improved capacity to disseminate advertising messages to migrants and expand their client base. While mobile technology generates substantial logistical and operational improvements for smuggling businesses, several core elements of the business remain unaffected. Trust between actors does not appear to be significantly enhanced through the use of mobile technology.

The demand side of the smuggling business – that is, relations between migrants or potential migrants and smugglers – has been significantly affected by improvements in mobile technology. The development of social media platforms has improved informational movement – that is, they have increased outreach and availability of information regarding irregular mobility regimes. Messenger services such as WhatsApp and Telegram have supported confidential discussions between smugglers and migrants about routes, costs, and logistical arrangements. This has improved informational movement as well – messenger services support more tailored and appropriate information.

Smuggling organisations have significantly benefited from mobile technology. In order to operate, smugglers require location-free information and contact points, communication devices, safe and secure meeting places, transportation, and time. As mobile technology has improved, communication devices have become cheaper and more effective, particularly compared to previous devices used by smuggling networks (satellite phones). Improvements in communication devices, in turn, have enabled smugglers to more easily, cheaply, and securely organise meeting places, accommodation, and transport; this has generated significant time savings. Organisations can operate more cheaply and effectively, thus increasing the reach of irregular mobility regimes, due to mobile technology.

The supply side of migrant smuggling networks relies on a network of contacts – in particular, contacts with other smuggling networks to permit movement to more distant geographic areas and contacts with authorities to enable smooth passage through borders. Mobile technology improves the links between different smuggling networks – but has not, according to the data gathered for this research, increased the trust between the networks. Mobile technology has not affected relationships with authorities but has increased risk of exposure. Although smugglers are aware of this risk, they have limited means to put in place mitigation measures.

This chapter represents only one case study regarding smugglers' perspectives on mobile technology in regard to their business. Authors including Sanchez (2017) rightly point out that many smuggling operations continue to use very basic technology, including non-internet enabled mobile phones, and acknowledge that lack of usage is more likely to be due to lack of available and affordable technology rather than lack of interest or sophistication on the part of smugglers (Sanchez, 2017). As mobile coverage increases and the cost of both smartphones and airtime decreases, it is likely that smuggling networks will make increasing use of this technology. The effects of mobile technology on smuggling are therefore likely to amplify, and the parallel mobility regime that has already started to develop is likely to institutionalise.

There are currently significant gaps in the literature around the ways in which mobile technology affect the smuggling business. As a result, the ways in which the smuggling business will expand, and the consequences of this expansion to migrants, as mobile technology becomes more affordable are not well understood. This research advocates for further effort to document smuggler narratives and experiences, particularly small-scale operations or components of operations with closer community ties. In order to better preserve the safety of migrants along the route, it is suggested that a series of parallel case studies across migration routes with greatest traffic, or in some of the most frequented smuggling hubs, is recommended. Such case studies would help to generate empirical evidence that can support more effective policy making, which in turn can help develop a more protective environment for migrants.

# NOTES

1. Irregular migration takes place when people move outside the regulatory norms of the origin, transit, and receiving countries (UNODC, 2018). Irregular migration consists of mixed migrants – asylum seekers, those seeking economic or educational opportunities, and victims of trafficking – with intentions and the nature of their movement changing along the route (IOM, 2016, accessed 2019).
2. This chapter acknowledges that there remains a significant amount of interaction between migrants and smugglers that continues to take place in person, due both to preference (Sanchez et al., 2018) and lack of appropriate communication technology.
3. Migrant smuggling is distinct from human trafficking in that smuggling is voluntary, though it is acknowledged that individual migrants may find that the voluntary and involuntary nature of their movement change along their route.
4. Mobile technology includes both hardware (mobile phones) and software (e.g., smartphone applications). Social media is defined broadly, and includes social networking sites (e.g., Facebook, Instagram), private messaging applications (e.g., Skype, WhatsApp, Viber, Telegram), web logs (blogs), and content communities (e.g., Wikipedia) (McGregor and Siegel, 2013; Frouws et al., 2016).

# REFERENCES

Bergman, M.M., Ohnmacht, T. and Maksim, H. (2009). *Mobilities and Inequality*. Ashgate, Bodmin.

Borkert, N., Fisher, K.E. and Yafi, E. (2018). The best, the worst and hardest to find: how people, mobiles, and social media connect migrants in(to) Europe. *Social Media + Society*. doi: 10.1177/2056305118764428

Bourdieu, P. and Wacquant, L. (1992). An invitation to reflexive sociology. 1st ed. University of Chicago Press, Chicago.

Brenner, Y and Frouws, B. (2019). Hype or hope? Evidence on use of smartphones & social media in mixed migration. Mixed Migration Centre, Geneva. Available from: http://www.mixedmigration.org/articles/hype-or-hope-new-evidence-on-the-use-of-smartphones-and-social-media-in-mixed-migration/ (last accessed 5 November 2019).

Broeders, D. and Engbersen, G. (2007). The fight against illegal migration: identification policies and immigrants' counterstrategies. *American Behavioral Scientist*, 50(12): 1592–609. https://doi.org/10.1177/0002764207302470

Campana, P. (2016). Explaining criminal networks: strategies and potential pitfalls. *Methodological Innovations*. January. doi:10.1177/2059799115622748

Crawley, H., Duvell, F., Jones, K. and Skleparis, D. (2016a). Understanding the dynamics of migration to Greece and the EU: drivers, decisions and destination. MEDMIG Research Brief No. 2. Available from: http://www.medmig

.info/research-brief-02-Understanding-the-dynamics-of-migration-to-Greece -and-theEU.pdf (last accessed 5 November 2019).

Crawley, H., Duvell, F., Jones, K., McMachon, S., and Sigona, N. (2016b). Destination Europe: understanding the dynamics and drivers of Mediterranean migration in 2015. MEDMIG Final Report. Available from: www.medmig .info/research-brief-destination-europe.pdf (last accessed 5 November 2019).

Dekker, R. and Engbersen, G. (2012). How social media transform migrant networks and facilitate migration. International Migration Institute Working Papers, No. 64. University of Oxford, Oxford.

Dekker, R. and Engbersen, G. (2014). How social media transform migrant networks and facilitate migration. *Global Networks*, 14: 401–18.

Dekker, R., Engbersen, G., Klaver, J. and Vonk, H. (2018). Smart refugees: how Syrian asylum migrants use social media information in migration decision-making. *Social Media + Society*. https://doi.org/10.1177/2056305118764439

EMSC (European Migrant Smuggling Centre) (2015). A study on the smuggling of migrants. EMSC, Brussels. https://ec.europa.eu/home-affairs/sites/homeaffairs/files/what-we-do/networks/european_migration_network/reports/docs/emn-studies/study_on_smuggling_of_migrants_final_report _master_091115_final_pdf.pdf (last accessed 5 November 2019).

EMSC (European Migrant Smuggling Centre) (2018). Two Years of EMSC: Activity Report January 2017–January 2018. Europol, Brussels.

Frouws, B., Phillips, M., Hassan, A. and Twigt, M. (2016). Getting to Europe the Whatsapp Way: the use of ICT in contemporary mixed migration flows to Europe. Regional Mixed Migration Secretariat Briefing Paper, Nairobi. Available from: https://papers.ssrn.com/sol3/papers.cfm?abstract_id =2862592 (last accessed 5 November 2019).

Gillespie, M., Ampofo, L., Cheesman, M. et al. (2016). *Mapping Refugee Media Journeys: Smartphones and Social Media Networks*. The Open University – France Medias Monde, Paris.

Gillespie, M., Osseiran, S. and Cheesman, M. (2018). Syrian refugees and the digital passage to Europe: smartphone infrastructures and affordances. *Social Media + Society*. https://doi.org/10.1177/2056305118764440

Grabowicz, P.A., Ramasco, J.J., Gonçalves, B. and Eguíluz, V.M. (2014). Entangling mobility and interactions in social media. *PLoS One*, 9(3): e92196. https://doi.org/10.1371/journal.pone.0092196

Hacsek, Z. and Visnansky, B. (2017). The impact of social media on the smuggling of migrants. Regional Academy on the United Nations. Available from: http://www.ra-un.org/uploads/4/7/5/4/47544571/2_unodc_2_final_paper.pdf (last accessed 5 November 2019).

Haug, S. (2008). Migration networks and migration decision-making. *Journal of Ethnic and Migration Studies*, 34(4): 585–605.

IOM (International Organization for Migration) (2016). Migrant Smuggling Data and Research: A Global Review of the Emerging Evidence Base (ed. M. McAuliffe and F. Laczko). IOM, Geneva. Available from https:// publications .iom.int/system/files/smuggling_report.pdf

IOM (International Organization for Migration) (2018). https://www.iom.int/key-migration-terms (last accessed 5 November 2019).

Koser, K. (1997). Negotiating entry into fortress Europe: the migration strategies of asylum seekers. In P. Muus (ed.), *Exclusion and Inclusion of Refugees in Contemporary Europe*. ERCOMER, Utrecht, pp.157–70.

Koser, K. and Pinkerton, C. (2002). The Social Networks of Asylum Seekers and the Dissemination of Information about Countries of Asylum, Home Office, Research Development and Statistics Directorate, London.

Koslowski, R. (2011) Global mobility regimes: a conceptual framework. In *Global Mobility Regimes*. Palgrave Macmillan, New York.

Lin, N. (1999). Building a theory of social network capital. *Connections*, 22(1): 28–51.

McGregor, E. and Siegel, M. (2013). Social media and migration research. MERIT Working Papers 2013-068. United Nations University – Maastricht Economic and Social Research Institute on Innovation and Technology (MERIT).

Milivojević, S. (2018). 'Stealing the fire', 2.0 style? Technology, the pursuit of mobility, social memory and de-securitization of migration. *Theoretical Criminology*, 23(2): 211–27. https://doi.org/10.1177/1362480618806921

Mohammadi, A., Nimkar, R. and Savage, E. (2019). 'We are the ones they come to when nobody can help': Afghan smugglers' perceptions of themselves and their communities. Migration Research Series, No. 56. International Organization for Migration (IOM), Geneva.

Monsutti, A. (2008). Afghan migratory strategies and the three solutions to the refugee problem. *Refugee Survey Quarterly*, 27(1): 58–73.

Musto, J.L. and Boyd, D. (2014). The trafficking-technology nexus, social politics. *International Studies in Gender, State & Society*, 21(3): 461–83.

Newell, B.C., Gomez, R. and Guajardo, V.E. (2016.) Information seeking, technology use, and vulnerability among migrants at the United States–Mexico border. *The Information Society* 32(3):176–91. https://doi.org/10.1080/01972243.2016.1153013

O'Reilly, K. (2015). Migration theories: a critical overview. In A. Triandafyllidou (ed.), *Routledge Handbook of Immigration and Refugee Studies*. Routledge, Abingdon, Oxford, pp. 1–9.

REF (Research and Evidence Facility) (July 2017). Migration between the Horn of Africa and Yemen: a study of Puntland, Djibouti and Yemen, London and Nairobi. EU Trust Fund for Africa (Horn of Africa Window) Research and Evidence Facility. Available from: http://www.soas.ac.uk/hornresearch-ref (last accessed 5 November 2019).

Rettie, R. (2008). Mobile phones as network capital: facilitating connections. *Mobilities*, 3(2): 291–311. doi: 10.1080/17450100802095346

Roberts, Z. (2017). Dissertation: information exchange between smugglers and migrants: an analysis of online interaction in Facebook groups. Cambridge University Press, Cambridge.

RVI (Rift Valley Institute) (2016). Going on Tahriib: the causes and consequences of Somali youth migration to Europe. Available from: https://www.refworld.org/docid/57e92d114.html (last accessed 5 November 2019).

Sanchez, G. (2017). Critical perspectives on clandestine migration facilitation: an overview of migrant smuggling research. *Journal of Migration and Human Security*, 5(1): 9–27.

Sanchez, G., Hoxhaj, R., Nardin, S., Geddes, A., Achilli, L. and Kalantaryan, R.S. (2018). A study of the communication channels used by migrants and asylum seekers in Italy, with a particular focus on online and social media. European Commission, Brussels.

Stone-Cadena, V. and Velasco, S.A. (2018). Historicizing mobility: Coyoterismo in the indigenous Ecuadorian migration industry. *The Annals of the American Academy of Political and Social Science*, 676(1): 194–211. doi:10.1177/0002716217752333

Tinti, P. and Reitano, T. (2017). *Migrant, Refugee, Smuggler, Savior*. Oxford University Press, Oxford.

Triandafyllidou, A. and Maroukis, T. (2012). *Migrant Smuggling: Irregular Migration from Asia and Africa to Europe*. Palgrave Macmillan, London.

UNODC (United Nations Office on Drugs and Crime) (2018). *Global Study on Smuggling of Migrants*. United Nations Publication, No. E.18. IV.9.

Urry, J. (2007). *Mobilities*. Polity, Cambridge.

Urry, J. (2012). Social networks, mobile lives and social inequalities. *Journal of Transport Geography*, 21, 24–30. doi:10.1016/j.jtrangeo.2011.10.003

van Liempt, I. (2007). *Navigating Borders: Inside Perspectives on the Process of Human Smuggling into the Netherlands*. Amsterdam University Press, Amsterdam.

Zhang, S.X., Chin, K.-L. and Miller, J. (2007). Women's participation in Chinese transnational smuggling: a gendered market perspective. *Criminology* 45(3): 699–733.

Zijlstra, J. and van Liempt, I. (2017). Smart(phone) travelling: understanding the use and impact of mobile technology on irregular migration journeys. *International Journal of Migration and Border Studies*, 3(2/3):174–91.

# 7. What shapes the attitude of the European Parliament voters toward migration? A comparative case study on Finland, Hungary and Bulgaria

**Deniz Yetkin Aker**

## INTRODUCTION

Since 1945, migration toward Europe has continued to increase with differences in countries in the region (Hansen, 2003, p. 25). Based on interviews with more than 183,000 adults across over 140 countries (between the years of 2012 and 2014), a project called *How the World Views Migration*[1] shows, for the first time, global public attitudes toward immigration (Esipova et al., 2015, p. 1). Most importantly, it shows that "in every major region of the world – with the important exception of Europe – people are more likely to want immigration levels in their countries to either stay at the present level or to increase, rather than to decrease" (Esipova et al., 2015, p. 1). The report indicates that European residents are, in general, "the most negative globally towards immigration, with the majority believing immigration levels should be decreased" (Esipova et al., 2015, p. 1).

In addition, according to a recent study on refugees' postings on Facebook about the countries hosting them ("the Digital Refuge"), the highest negative Facebook sentiments of refugees toward European Union (EU) member states are seen in Finland, Bulgaria, and Hungary (https://digitalrefuge.berkeley.edu). Although European residents have in general the most negative attitude toward immigration worldwide, the main similarity among these countries in Europe is the highest negative sentiments of immigrants (mainly refugees) toward these host countries. There can be many reasons such as a non-welcoming society, politics, policies and/or social context that are reinforced, for instance, through media and the internet. Within this framework, this study aims to focus mainly on the public of Finland, Hungary and Bulgaria to design a case study and aims to understand the reasons for the negative attitudes of these host countries' voters who voted in the European Parliament (EP) elections in 2019.[2] It is expected that there is a close and direct connection between their negative attitudes toward immigration with life satisfaction, political behavior of voters, social media and internet usage, their general ideas about their countries and their demographic profile.

The arguments and discussions are presented by, firstly, discussing the literature on attitudes toward immigrants. Secondly, the methods and data are explained in detail. Lastly, the chapter concludes with a discussion of policy implications.

## CONCEPTUAL FRAMEWORK AND HYPOTHESES

Studies focusing on individuals' attitudes toward migration explore several factors such as the role of social context, contact patterns of individuals (Hood and Morris, 1997, in Neiman et al. 2006, p. 36) and their economic situations (Citrin et al., 1997, in Neiman et al., 2006, p. 36), xenophobia, voters' economic interests (O'Rourke and Sinnott, 2006, p. 839), as well as immigrants' country of origin (Nonchev et al., 2012, p. 77). Mayda (2006), for instance, focuses on the economic and non-economic reasons of individual attitudes toward immigrants. By using two survey data sets, she argues that there is a correlation between individuals' skills and their pro-immigration attitudes. Thus, if natives are more skilled than immigrants, then they have pro-immigrant attitudes (supporting immigration etc.). However, they are opposed to immigration, if it is otherwise (Mayda, 2006, p. 511).

Different from the studies on individuals' attitudes toward migration, this study aims to discuss mainly voters' attitudes in a comparative manner: it aims to analyse whether demographics, life satisfaction, internet as well as factors related to political behavior of individuals (such as political interest) have any effect on their attitude toward immigration in Finland, Hungary and Bulgaria.

Firstly, in democracies, voting is used as a conventional method to influence policies (van der Brug et al., 2000, p. 79). Focusing on the EP election 2019, it is seen that with 50.6%, overall turnout was the highest since 1994. According to the post-electoral Eurobarometer survey results, the turnout increase can be explained mostly by the youth: "19 Member States registered increases in voter turnout since 2014, especially Poland, Romania, Spain, Austria, Hungary and Germany as well as Slovakia and Czechia, where turnout is traditionally very low" (2019 EP Election, Press Releases). Without discussing why voters in Finland, Bulgaria or Hungary vote for different parties in their countries, this study aims to understand whether voting for different parties is related to individuals' attitudes toward migration.[3]

The first hypothesis is as follows:

*H1: Voters' attitude scores toward immigration – who voted for different parties – are significantly different for Finnish, Hungarian and Bulgarian cases.*

Thus, it is hypothesized that voters supporting different parties in EP elections in 2019 have different attitudes toward immigrants in Finland, Hungary and Bulgaria.

Secondly, several studies support the idea that while there is a "decline in long-term influences on the vote" the potential for issue voting has been increasing (Dalton, 2000, p. 924). Among the issues existent in the data set, immigration is selected as an issue for this study. The reason is that, for instance, in Finland since 2008, immigration and multiculturalism have been frequently discussed in media and politics. There has been a consensus about a strict immigration policy in Finland for several decades. In 2008, after the municipal elections, "anti-immigration issues became the focus of broad interest in media and politics" (Keskinen et al., 2009, in Keskinen, 2013, p. 227). Similarly, in Hungary, several studies show that Muslim migrants especially were "exploited by the governing populist radical right party Fidesz and the right-wing extremist party Jobbik to re-narrate the enemies of the Hungarian nation" (Thorleifsson, 2017, p. 319). As another example, for more than 30 years, Bulgaria has been a sending country. Although in 2012 "the main concerns and fears of Bulgarians were the internal problems, low salaries, unemployment, corruption," in particular since 2017 the public has been in fear of terrorism and the refugee crisis and the issue of migration became politicized. According to "a recent international study by WIN/Gallup International association on public opinion about migration," it is suggested that Bulgarians have a very negative opinion about immigration (Bobeva, 2017, p. 40).

Overall, it is expected that voters could be motivated and affected by the issue of migration. This study aims to analyse whether voters, who are motivated by the issue of immigration (have sensitivity to immigration), have different attitudes toward migration than others motivated by other issues such as combatting climate change. The second hypothesis is as follows:

*H2: It is expected that if voters are voting because of migration, then this situation will predict their attitude toward immigration in Finland, Hungary and Bulgaria.*

Several studies such as those by Verba and his colleagues (Verba et al., 1978; 1995, in Dalton, 2000, p. 927) have verified that different forms of action have different political implications. In democracies, the public is expected to be involved and influence policies by taking an interest in politics and engaging in political discussion (Dalton, 2000, p. 927). In line with these arguments, it is expected that having a general interest in politics (having discussions with friends and relatives about political matters) could affect their attitude toward immigration.

*H3: Voters' interest in politics (having political discussions) predicts their attitude toward immigration in Finland, Hungary and Bulgaria.*

The internet offers a new medium for the rapid distribution of information and communication which is cheap, fast, geographically unbound, and accepted as free and egalitarian "in that hardly any central control exists to filter out specific content" (Albrecht, 2006, p. 64).

As an example, by widespread usage of social media, individuals engage with and share news and political information (Kümpel et al., 2015; Weeks and Holbert, 2013, in Weeks et al., 2017, p. 363). Many individuals live a greater part of their social, professional and political lives online on, for instance, Facebook and Twitter. They share their arguments and views on these platforms. As an example, Europeans spend around four hours per day on the internet and about three in four Europeans use a minimum of one social network website (Bartlett, 2014, p. 100). It is also possible that "populist parties in Europe have been quicker to spot the opportunities these new technologies present to reach out and mobilize an increasingly disenchanted electorate" (Bartlett, 2014, p. 100). In addition, the internet (social media, blogs etc.) can reinforce existing problems (Albrecht, 2006, p. 64) and can create fragments in political discourse (Papacharissi, 2002, p. 9). The fragments could be related to international migrants and political information and/or disinformation about international migration, and these could be distributed through the internet and/or social media. In line with these arguments, to measure internet and social media effect on individuals, the question about internet usage is selected as a variable. It is expected that using the internet in general for at least 15 days before elections could affect voters' attitude toward immigration in Finland, Hungary and Bulgaria.

*H4: Voters' attitude toward migration is correlated with voters' internet usage in general (15 days) before EP elections in Finland, Hungary and Bulgaria.*

*H5: Voters' internet usage rates (15 days) before EP elections predict voters' attitude toward immigration in the cases of Finland, Hungary and Bulgaria.*

In addition to these possible factors, general life satisfaction in one's own country and an individual's ideas about the country's general situation can affect an individual's attitude toward migration. As an example, in the *How the World Views Migration*, scholars indicate that "those who perceive economic situations as poor or worsening are more likely to favor lower immigration levels into their countries" (Esipova et al., 2015, p. 1). The sixth hypothesis is as follows:

*H6: It is expected that life satisfaction rates of voters and voters' ideas about their country's situation in general predict their attitude toward immigration in Finland, Hungary and Bulgaria.*

Lastly, since women and men may have different patterns of judgment which are shaped by their social status (Kalaycioğlu, 1983, p. 19), gender difference can affect their attitude toward immigration. Individuals are politically socialized at different periods in time; thus, the ideas and political attitudes of older or younger individuals could differ from each other (Kalaycioğlu, 1983, p. 20). In this study, it is expected that different age categories could affect voters' attitudes toward immigration differently. The political ideology is another factor that can shape individuals' attitude toward important issues such as international migration. It is also expected that left-right ideological self-placement influence individuals' attitude toward immigration; since, as several studies show, the "positions of voters as well as parties on a variety of different issues are largely structured by left/right ideology" (e.g. Fuchs and Klingemann 1990; Klingemann et al., 1994; Van der Brug, 1997; van der Eijk and Niemöller, 1983, in Van der Brug et al., 2000, p. 80).

Other hypotheses of this study are as follows:

*H7: It is expected that the means of attitude scores toward immigration of men and women voters are significantly different in Finland, Hungary and Bulgaria.*

*H8: Voters' self-placements on the left-right ideological dimension and voters' political interest predict their attitude toward immigration in Finland, Hungary and Bulgaria.*

*H9: Voters' age differences predict their attitude toward immigration in Finland, Hungary and Bulgaria.*

In the following section, the methods and data are explained for critical analysis of the results from the study.

## DATA AND METHODS

Since the 1970s, the European Commission's Standard Eurobarometer has been regularly conducted in EU member countries to monitor public opinion on several issues such as financial crisis and related EU policies, attitudes toward the EU and European citizenship. The data used for this study, Eurobarometer 91.5 Data, was collected by the Commission between June and July 2019 as a face-to-face survey in EU member states and candidates. Its questionnaire covers key areas such as national employment situation, knowledge of and trust in selected institutions, media monitoring and voting behavior related to EP elections in 2019. The study sample includes 32,524 respondents and 872 variables. The study population covers residents in each of the 28 EU member states who are 15 years or older. In other countries such as Turkey and North Macedonia, the study population includes "the national population of citizens and the population of citizens of all the European Union Member States that are residents in these countries and have a sufficient command of the national languages to answer the questionnaire" (Eurobarometer 91.5 Data).

By using the Digital Refugee project results, this study aims to compare the EU member states where negative Facebook sentiments of refugees remain the highest. To explore the European Refugee Crisis, "The Digital Refugee Project" was designed and a website created, called "Digital Refuge." The Project mainly focuses on the height of the crisis, between the years of 2011 and 2017. To develop it, several data sources were used such as official reports and data published by the United Nations High Commission for Refugees (UNHCR) and the EU. In addition, they analysed "over 10,900 public social media posts, primarily in Arabic and Farsi, and hundreds of thousands of reactions to these posts" (The Digital Refuge website). The posts were collected from "public Facebook groups frequently used by asylum seekers to access information over the course of the crisis" (The Digital Refuge website). They included "over 6,000 in-person interviews conducted across Greek refugee settlements, formal and informal, by the NGO Internews" to their social media analysis. Digital Refugee aims to map the experiences of refugees in their own words (The Digital Refuge website).

The Project results are explored through interactive bar charts, where, for instance, the degree to which country level of Facebook posting themes and sentiments can be compared. It should be emphasized that although the project monitors and analyses 10,951 posts, since most of these posts have been widely shared and followed (average 80% reactions per post) it is possible for the study to capture "widely shared sentiments and a wide discussion within the refugee community" (The Digital Refuge website).

After monitoring 10,951 Facebook posts of the refugees, the project results suggest that "the refugee crisis is quite global" and some themes and sentiments are specific to countries (The Digital Refuge website). The interactive bar charts allow a country-level comparison of public social media posts (Facebook posts) themes and sentiments. There are 41 countries with 15 or more posts in these charts. When refugees' Facebook posts mention a country, the general sentiments range from 1 (negative) to 3 (positive). For instance, among the collected posts, 1,234 relate to Australia with a score of 1.35 (very negative). For Australia, 17 of the posts are related to education (in a negative way) and 27 are related to healthcare (in a negative way). As another example, among the collected posts, 1,116 are related to Germany with a score of 1.92

(positive/center). For Germany, 364 of the posts are related to asylum policy (in a positive/center way).

For this study, EU member states with the highest negative sentiment scores are selected from Eurobarometer Data. Thus, in Digital Refuge data, since there are three EU member states with the highest negative general sentiment scores,[4] this study focuses only on these three countries (Finland, Hungary and Bulgaria) as the case study and their data is drawn from the Eurobarometer 91.5 data set. To understand how voters in these receiving EU countries with the highest negative sentiment scores feel about international migration, the main aim of this study is to critically analyse factors that shape the attitude of Finnish, Hungarian and Bulgarian voters – who voted in EP elections in 2019 – toward migration.

The population of this study includes only the Finnish, Hungarian and Bulgarian public and the sample consists of 3,157 respondents (1,023 Finnish; 1,069 Hungarian; 1,065 Bulgarian participants). Finnish, Hungarian and Bulgarian voters (who voted in the 2019 EP elections) were filtered from the data. After filtering out the other country cases, the new sample size of 1,724 respondents includes 567 Finnish, 565 Bulgarian, and 592 Hungarian participants.[5]

## OPERATIONALIZATION OF VARIABLES

### Dependent Variable

In the literature, to measure the attitudes of individuals toward migration-related topics, several questions have been used such as "whether they think levels of immigration should increase, remain the same, or be reduced" (Simon and Lynch, 1999, in Neiman et al., 2006, p. 36). In addition, several assessments have been used such as the contributions immigrants made to … (host countries) that can be rated on "5-point *Agree–Disagree* scales with higher scores indicating more positive attitudes toward immigrants" (Ward and Masgoret, 2008, p. 232).

In this study, the dependent variable which measures the attitude of voters toward international migration – who voted in the 2019 EP elections – is a continual variable ranging from 1 to 4 and created as a result of factor analysis of four different variables. To operationalize the dependent variable and to measure various facets of voters' attitudes, several questions and statements were selected from the questionnaire.

The statements are about individuals' ideas about migration of people from EU member states and from outside the EU, whether immigrants contribute to the participants' country, and whether their country should help refugees.

The selected statements are:

- Please tell me whether each of the following statements evokes a positive or negative feeling for you:
  - Immigration of people from other EU Member States
  - Immigration of people from outside the EU (Eurobarometer 91.5, QB3)
- To what extent do you agree or disagree with each of the following statements?:
  - Immigrants contribute a lot to (OUR COUNTRY)
  - (OUR COUNTRY) should help refugees (Eurobarometer 91.5, QD9)

The first question ranges from 1 (very positive) to 4 (very negative) and 5 (Don't Know). The second question ranges from 1 (totally agree) to 4 (totally disagree) and 5 (Don't Know). The answers are recoded as they range from 1 to 4 by eliminating "Don't Know" answers. Such recoding methods might decrease the number of observations; however, it results in a stronger analysis.

Exploratory factor analysis (EFT) was conducted to refine the list of these statements and group similar ones into "attitude toward migration" to verify scale construction. With such a scale it is aimed to show voters' attitude (like being positive or agree upon vs being negative) toward international migration in general. The results show that it is possible to construct a scale by using these selected variables to measure voters' attitude toward migration.[6]

The compute dialog box was used to measure the means over the four variables that are about individuals' ideas about migration of people from EU member states and from outside the EU, whether immigrants contribute to the participants' country, and whether their country should help refugees. As a result, a new scale variable (the dependent variable) – average score of voters' attitudes toward international migration – was created (1 to 4, from positive to negative). The Independent Samples *t* Test is used to test statistical differences between the means of attitude scores of Finnish, Hungarian and Bulgarian voters toward immigration.

The alternative hypothesis ($H_1$) for each of the Independent Samples *t* Test can be expressed as:

*There is statistical evidence that the means of attitude scores of Finnish, Hungarian and Bulgarian voters are significantly different from each other.*

The tests show that the null hypotheses can be rejected; thus, there are significant differences in the means of attitude scores between the voters. The mean of the attitude score of all voters from the sample is 2.5. Comparing the group (Bulgaria and Hungary) with Finland, it is realized that the attitudes of Finnish voters toward immigration are different from other voters (for Finland, $t_{13.359} = 2.21$, $p < .000$, for others $t_{13.359} = 2.64$, $p < .000$). Their attitude is more positive than Hungarians and Bulgarians. The mean of Hungarian voters' attitude scores is 2.6 and the mean of Bulgarian voters' attitude scores is 2.7. Bulgarian voters' attitude seems to be much more negative toward migration. This study attempts to understand such variations of voters' attitudes in each of these countries.

## Independent Variables

As a result of the literature review, the study includes mainly eight independent variables as possible predictors. These are life satisfaction, political interest, reason to vote (issue cause to vote) such as immigration, voters' internet usage rates (15 days) before EP elections, EP Election-selected Parties and demographic profile.

In the questionnaire, life satisfaction scale ranges from 1 (very satisfied) to 4 (not at all satisfied) and 5 (Don't Know (DK)), which is recoded into four brackets by removing the "Don't Know" answer. To measure political interest, the summary of two questions is used as a proxy: the questions asked, for instance, whether the participants discuss national political matters and local political matters with their friends or relatives. The scale includes four brackets: "strong, medium, low and not at all" ("Don't Know" answer is removed). Reason to vote (issue cause to vote) category is a dummy variable (others as 0 and immigration as 1). Voters' internet usage (15 days) before EP elections is drawn from respondents' self-reports to measure their internet usage frequency. It includes six brackets: "everyday/almost every day, twice or three times a week, about once a week, less often, never, No access to this medium and DK." The answer "Don't Know" is removed and the variable is recoded. EP Election-Parties category is drawn from respondents' self-reports to understand the vote share of parties running in the 2019 EP elections in each of the cases (different parties for three different cases) (Eurobarometer 91.5 Data).

Besides these independent variables, looking at demographics can also be important for the model. The selected demographics are gender, age and left-right ideological self-placement. Among the demographic variables, the age band is recoded into four brackets (15–24, 25–39, 40–54, and 55 and older). Other demographic variables are gender (recoded as a dummy variable, man as 0 and woman as 1) and left-right ideological dimension. The left-right ideological self-placement is coded into five brackets by removing the "Don't Know" answer (Eurobarometer 91.5 Data).

### Analytical Technique

By using SPSS, the statistical analysis software, descriptive statistics are conducted to summarize the sample characteristics. In addition, Pearson correlation coefficients are used to examine bivariate relationships among several variables. After *t* Test and Variance analysis, multiple regression analyses are conducted for Hungary, Finland and Bulgaria. The analyses are made on an individual basis.

## RESULTS

### Descriptive Result

Table 7.1, the frequency distribution table, represents the number of observations (the frequency) and percentages for each unique value of the independent variables used in this study. As shown, other than in Bulgaria, respondents in Finland and Hungary are mostly satisfied with their lives in their countries. It is seen that almost half of the Bulgarian respondents are not satisfied with their lives.

Most of the respondents point out various reasons – other than immigration – to vote. Nevertheless, migration as a reason to vote is highest in the case of Hungary (18.4%). In addition, according to the results, in Finland and Bulgaria, most of the respondents place themselves in the center (38.3 and 23.9 respectively) but in Hungary on the right (27.7%). The highest percentage for the left is seen in the case of Bulgaria (19.3%).

*Table 7.1* *Sum of descriptive statistics*

| *Variables/Cases* | *Finland N = 567* | | *Hungary N = 592* | | *Bulgaria N = 565* | |
|---|---|---|---|---|---|---|
| **Life Satisfaction** | Frequency | Percent | Frequency | Percent | Frequency | Percent |
| *Very satisfied* | 232 | 40.9 | 77 | 13.0 | 52 | 9.2 |
| *Fairly satisfied* | 317 | 55.9 | 404 | 68.2 | 257 | 45.5 |
| *Not very satisfied* | 17 | 3.0 | 94 | 15.9 | 185 | 32.7 |
| *Not at all satisfied* | 1 | .2 | 17 | 2.9 | 67 | 11.9 |
| *Total* | 567 | 100.0 | 592 | 100.0 | 561 | 99.3 |
| *Missing* | – | – | – | – | 4 | .7 |
| *Total* | 567 | 100.0 | 592 | 100.0 | 565 | 100.0 |
| **Reason to Vote (issue cause to vote) – Immigration** | Frequency | Percent | Frequency | Percent | Frequency | Percent |
| *Other Reasons* | 500 | 88.2 | 477 | 80.6 | 505 | 89.4 |
| *Immigration* | 38 | 6.7 | 109 | 18.4 | 27 | 4.8 |
| *Total* | 538 | 94.9 | 586 | 99.0 | 532 | 94.2 |
| *Missing* | 29 | 5.1 | 6 | 1.0 | 33 | 5.8 |
| *Total* | 567 | 100.0 | 592 | 100.0 | 565 | 100.0 |
| **Gender** | Frequency | Percent | Frequency | Percent | Frequency | Percent |
| *Man* | 276 | 48.7 | 278 | 47.0 | 260 | 46.0 |
| *Woman* | 291 | 51.3 | 314 | 53.0 | 305 | 54.0 |
| *Total* | 567 | 100.0 | 592 | 100.0 | 565 | 100.0 |
| **Age** | Frequency | Percent | Frequency | Percent | Frequency | Percent |
| *15–24 years* | 15 | 2.6 | 25 | 4.2 | 24 | 4.2 |
| *25–39 years* | 65 | 11.5 | 102 | 17.2 | 112 | 19.8 |
| *40–54 years* | 110 | 19.4 | 196 | 33.1 | 168 | 29.7 |
| *55 years and older* | 377 | 66.5 | 269 | 45.4 | 261 | 46.2 |
| *Total* | 567 | 100.0 | 592 | 100.0 | 565 | 100.0 |
| **Left-Right Political Dimension** | Frequency | Percent | Frequency | Percent | Frequency | Percent |
| *(1–2) Left* | 28 | 4.9 | 42 | 7.1 | 109 | 19.3 |
| *(3–4)* | 114 | 20.1 | 88 | 14.9 | 63 | 11.2 |
| *(5–6) Center* | 217 | 38.3 | 132 | 22.3 | 116 | 20.5 |
| *(7–8)* | 131 | 23.1 | 160 | 27.0 | 135 | 23.9 |
| *(9–10) Right* | 60 | 10.6 | 164 | 27.7 | 101 | 17.9 |
| *Total* | 550 | 97.0 | 586 | 99.0 | 524 | 92.7 |

| ***Variables/Cases*** | ***Finland N = 567*** | | ***Hungary N = 592*** | | ***Bulgaria N = 565*** | |
|---|---|---|---|---|---|---|
| *Missing* | 17 | 3.0 | 6 | 1.0 | 41 | 7.3 |
| *Total* | 567 | 100 | 592 | 100.0 | 565 | 100.0 |
| **Voters' internet usage (15 days) before EP elections** | Frequency | Percent | Frequency | Percent | Frequency | Percent |
| *Everyday/Almost everyday* | 217 | 38.3 | 200 | 33.8 | 268 | 47.4 |
| *Two or three times a week* | 54 | 9.5 | 104 | 17.6 | 65 | 11.5 |
| *About once a week* | 31 | 5.5 | 30 | 5.1 | 19 | 3.4 |
| *Less often* | 38 | 6.7 | 28 | 4.7 | 18 | 3.2 |
| *Never* | 197 | 34.7 | 207 | 35.0 | 164 | 29.0 |
| *No access to this medium* | 23 | 4.1 | 13 | 2.2 | 25 | 4.4 |
| *Total* | 560 | 98.8 | 582 | 98.3 | 559 | 98.9 |
| *Missing* | 7 | 1.2 | 10 | 1.7 | 6 | 1.1 |
| *Total* | 567 | 100.0 | 592 | 100.0 | 565 | 100.0 |
| ***EP Election-Party Voted For*** | Frequency | Percent | Frequency | Percent | Frequency | Percent |
| *Party 1* | 99 | 17.5 | 360 | 60.8 | 214 | 37.9 |
| *Party 2* | 124 | 21.9 | 38 | 6.4 | 172 | 30.4 |
| *Party 3* | 90 | 15.9 | 47 | 7.9 | 47 | 8.3 |
| *Party 4* | 36 | 6.3 | 12 | 2.0 | 41 | 7.3 |
| *Party 5* | 90 | 15.9 | 84 | 14.2 | 17 | 3.0 |
| *Party 6* | 15 | 2.6 | 36 | 6.1 | 24 | 4.2 |
| *Party 7* | 75 | 13.2 | 5 | .8 | 10 | 1.8 |
| *Party 8* | 26 | 4.6 | 9 | 1.5 | 3 | .5 |
| *Party 9* | 2 | .4 | – | – | 12 | 2.1 |
| *Party 10* | 3 | .5 | – | – | 6 | 1.1 |
| *Party 11* | 3 | .5 | 1 | .2 | 8 | 1.4 |
| *Party 12* | – | – | – | – | 4 | .7 |
| *Party 13* | – | – | – | – | 2 | .4 |
| *Party 14* | – | – | – | – | – | – |
| *Party 15* | – | – | – | – | 1 | .2 |
| *Other (SPONTANEOUS)* | 4 | .7 | – | – | 4 | .7 |
| *Total* | 567 | 100.0 | 592 | 100.0 | 565 | 100.0 |

| *Variables/Cases* | *Finland N = 567* | | *Hungary N = 592* | | *Bulgaria N = 565* | |
|---|---|---|---|---|---|---|
| ***Political Interest Index*** | Frequency | Percent | Frequency | Percent | Frequency | Percent |
| *Strong* | 125 | 22.0 | 107 | 18.1 | 99 | 17.5 |
| *Medium* | 324 | 57.1 | 379 | 64.0 | 339 | 60.0 |
| *Low* | 65 | 11.5 | 51 | 8.6 | 80 | 14.2 |
| *Not at all* | 53 | 9.3 | 55 | 9.3 | 47 | 8.3 |
| *Total* | 567 | 100.0 | 592 | 100.0 | 565 | 100.0 |

*Note:* The table is created using Eurobarometer 91.5 Data and 2019 European Election Results (accessed August 21, 2020 at https://www.europarl.europa.eu/election-results-2019/en).

The majority of the sample is female and most of the respondents are categorized in the age class "55 years and older." Interestingly, most of the respondents in each case report that they either use the internet every day before the EP election or never.

Focusing on the parties having more than 80 votes (post-election survey), as Table 7.2 shows, in Finland, although most of the respondents place themselves in the center, 18% of the respondents support KESK/EP Group-Renew Europe (Party 1), 22% support KOK/EP Group-Christian Democrats (Party 2) and 20% support either SDP (90)/EP Group-Socialists and Democrats (Party 3) or VIHR(90)/EP Group-Greens and EFA (Party 5). In Hungary, it is seen that 61% of the respondents support FIDEZS and KNDP/EP Group-Christian Democrats (Party 1), and 14.2% support DK/EP Group-Socialists and Democrats (Party 5). For Bulgaria, the table shows that 38% of the respondents support GERB/EP Group-Christian Democrats (Party 1) and 30.4% support BSP/EP Group-Socialists and Democrats (Party 2). For a detailed analysis, parametric tests and regression results are analysed and results provided in Tables 7.2 to 7.8.

## Parametric Tests – Testing Hypotheses

### *t* Test

The Independent Samples *t* Test is used to test statistical differences between the means of attitude scores toward immigration of men and women voters (gender variable) for Finnish, Hungarian and Bulgarian cases.

*Table 7.2 Parties in Finland (post-election survey order, parties with more than 80 votes)*

| | | | | |
|---|---|---|---|---|
| Finland | 1. KESK (99)/ EP Group-Renew Europe | 2. KOK(124)/EP Group-Christian Democrats | 3. SDP (90)/EP Group-Socialists and Democrats | 4. VIHR (90)/EP Group-Greens and EFA |

*Note:* The table is created using 2019 European Election Results (accessed August 21, 2020 at https://www.europarl.europa.eu/election-results-2019/en).

The alternative hypothesis ($H_1$) for each of the Independent Samples *t* Test can be expressed as:

*There is statistical evidence that the means of attitude scores of Finnish men and women voters toward immigration are significantly different.*

*There is statistical evidence that the means of attitude scores of Hungarian men and women voters toward immigration are significantly different.*

*There is statistical evidence that the means of attitude scores of Bulgarian men and women voters toward immigration are significantly different.*

For all of the cases, the null hypothesis cannot be rejected, thus, there is not any significant difference in means of attitude scores between men and women.[7] Rather than expected, attitudes of men and women voters toward immigration are not different in all the three cases. Thus, gender does not matter in voters' attitude differences.

## F Test – the Analysis of Variance

The Analysis of Variance (ANOVA) test is used to test whether there are statistical differences between the means of attitude scores toward immigration of Finnish voters – voting for different parties. The Hungarian and Bulgarian cases – since the data fails certain assumptions – are analysed using the Kruskal-Wallis H test (nonparametric alternative to the one-way ANOVA).

*Table 7.3* *Parties in Hungary and Bulgaria (post-election survey, parties with more than 80 votes)*

| | | |
|---|---|---|
| Hungary | 1. FIDEZS and KNDP (360)/EP Group-Christian Democrats | 2. DK (84)/EP Group-Socialists and Democrats |
| Bulgaria | 1. GERB (214)/EP Group-Christian Democrats | 2. BSP (172)/EP Group-Socialists and Democrats |

*Note:* The table is created using 2019 European Election Results (accessed August 21, 2020 at https://www.europarl.europa.eu/election-results-2019/en).

The alternative hypothesis ($H_1$) for F test can be expressed as:

*There is statistical evidence that the means of attitude scores toward immigration of Finnish voters who voted for different parties are significantly different.*

Focusing on the parties having more than 80 votes (see Table 7.2), in the Finland case, it is seen that there is a statistically significant difference between groups (voters who voted for different parties) as determined by one-way ANOVA.[8] A (Tukey HSD) post hoc test reveals that voters' average attitude scores toward immigration are statistically significantly higher for the ones who voted for VIHR than the ones who support KESK. The test supports the hypothesis and shows that VIHR (EP group as Greens and EFA) voters seem to have more negative attitudes toward immigrants then the ones who support KESK (EP Group-Renew Europe, a liberal group).

**Nonparametric Tests**

The alternative hypothesis ($H_1$) for each Kruskal-Wallis H test (nonparametric alternative to the one-way ANOVA) can be expressed as:

*There is statistical evidence that the means of attitude scores toward immigration of Hungarian voters who voted for different parties are significantly different.*

*There is statistical evidence that the means of attitude scores toward immigration of Bulgarian voters who voted for different parties are significantly different.*

*Table 7.4* *Pearson correlations among several variables for the Finnish case*

| | | | | |
|---|---|---|---|---|
| *Attitude toward immigration* | – | | | |
| *Left-Right political dimension* | .174** | – | | |
| *Age* | .081 | .075 | – | |
| *Voters' internet usage (15 days) before EP elections* | .025 | .043 | .447** | – |
| *Political interest* | .108* | .018 | .035 | .097* |

*Note:* * Correlation is significant at the .05 level (2-tailed); ** correlation is significant at the .01 level (2-tailed).

In the Hungarian case, a Kruskal-Wallis H test shows that there is a statistically significant difference in attitude score of different groups of voters.[9] For the groups, see Table 7.3. Similarly, in the Bulgarian case, there is a statistically significant difference in attitude score of different groups of voters.[10] For the groups, see Table 7.4.

Parties could shape citizens' ideas "by mobilizing, influencing, and structuring choices among political alternatives" (Leeper and Slothuus, 2014, p. 129). According to these three hypotheses test results, in all of the three cases, different party supporters have different attitudes toward immigration.

## Correlations Results

The bivariate correlation results of Finland indicate that voters' attitude toward migration is positively correlated with the left-right political spectrum (a two-dimensional model, r = 0.174, a weak correlation) and political interest of voters (r = 0.108, a weak correlation). This means that in Finland while voters place themselves on the right, or while their interest in politics decreases, their attitude toward migration becomes more negative (1 very positive to 4 very negative). The strongest correlation is between voters' internet usage rates and age: r = 0.447 (a moderate correlation).[11]

*Table 7.5 Pearson correlations among several variables for the Hungarian case*

| | | | | |
|---|---|---|---|---|
| *Average attitude toward immigration* | – | | | |
| *Left-Right political dimension* | .174** | – | | |
| *Age* | .049 | -.183** | – | |
| *Voters' internet usage* | -.120** | -.003 | .373** | – |
| *Political interest* | .179** | -.013 | -.041 | -.031 |

*Note:* * Correlation is significant at the .05 level (2-tailed); ** correlation is significant at the .01 level (2-tailed).

Secondly, the bivariate correlation results of Hungary indicate that voters' attitude toward migration is positively correlated with the left-right political spectrum (r = 0.174, a weak correlation) and political interest of voters (r = 0.179, a weak correlation), while it is correlated with voters' internet usage rates negatively (r = -0.120, a weak correlation). That is, there is a relationship between internet usage, placing oneself on the right (right-wing) as well as political apathy and negative attitude toward immigration. Different from the Bulgarian case, for instance, while voters' internet usage increases (1 to 4 from every day to no access to the medium), their attitude toward migration becomes more negative (1 to 4 from very positive to very negative). While voters place themselves on the right of the left-right political spectrum, their attitude toward migration becomes more negative (1 very positive to 4 very negative). The strongest correlation is between voters' internet usage rates and age: r = 0.373 (a moderate correlation).[12]

Lastly, bivariate correlation results of Bulgaria indicate that voters' attitude toward migration is correlated with the left-right political spectrum (r = -0.226, a weak correlation), with voters' internet usage rates and with age of voters (r = 0.108, a weak correlation) positively. These results indicate that, in Bulgaria, as voters become older, their attitudes toward migration becomes more negative (1 very positive to 4 very negative). The results support the *How the World Views Migration* arguments claiming that "younger people generally tend to be more

*Table 7.6 Pearson correlations among several variables for the Bulgarian case*

| | | | | |
|---|---|---|---|---|
| *Average attitude toward immigration* | – | | | |
| *Left-Right political dimension* | -.226** | – | | |
| *Age* | .128** | -.378** | – | |
| *Voters' internet usage (15 days) before EP elections* | .091* | -.425** | .560** | – |
| *Political interest* | .004 | -.062 | -.098* | .049 |

*Note:* * Correlation is significant at the .05 level (2-tailed); ** correlation is significant at the .01 level (2-tailed).

positive towards immigration" (Esipova et al., 2015, p. 1). In addition, while they place themselves on the right scale of the political spectrum, their attitudes toward migration become more negative. Different from the Finnish case, while internet usage of Bulgarian voters decreases, their attitudes toward migration become more negative (1 very positive to 4 very negative). The strongest correlation is between voters' internet usage rates and age: r = 0.560 (a moderate correlation).[13]

## Multiple Regression Results

For a detailed analysis and to understand the causal relationship between voters' attitude toward immigrants in Finland, Bulgaria and Hungary and possible predictors (significant ones), multiple regression analysis for each case is conducted. It is hypothesized that the combination of life satisfaction of voters, their ideas about their country's situation in general, their internet usages (internet and/or social media effect), the left-right ideological spectrum, their interest in politics (having political discussions), the reason they vote (because of migration) and age differences can predict voters' attitude toward immigration in Finland, Hungary and Bulgaria. The results support the *How the World Views Migration* arguments claiming that "those who perceive economic situ-

*Table 7.7* *Multiple regression analysis for the Finnish case*

| ***Finland-Coefficients*** | | | | | | |
|---|---|---|---|---|---|---|
| | ***Model*** | | Unstandardized Coefficients | | Standardized Coefficients | Sig. |
| | | | B | Std. Error | Beta | |
| | *1* | **(Constant)** | 1.150 | .162 | | .000 |
| | | Life-Satisfaction (DK out)-recoded | .049 | .039 | .052 | .207 |
| | | **The situation in (our) country general-recoded** | .186 | .049 | .156 | .000 |
| | | Voters' internet usage (15 days) before EP elections-recoded | -.003 | .013 | -.012 | .792 |
| | | **Left-Right Political Dimension (5-scale)-recoded** | .074 | .021 | .144 | .000 |
| | | Age (Recoded 4 Categories) | .056 | .030 | .085 | .062 |
| | | Political Interest Index (D71 SUMMARIZED) | .067 | .026 | .106 | .009 |
| | | **Reason to Vote (issue cause to vote) – Immigration** | .672 | .085 | .321 | .000 |

*Note:* Dependent variable: average attitude scale toward immigration.

ations as poor or worsening are more likely to favor lower immigration levels into their countries" (Esipova et al., 2015, p. 1).

Firstly, in the Finnish case, the multiple regression model (simultaneous regression) with all ten predictors produces an adjusted $R^2 = .174$, $F (7, 502) = 16.354$, $p < .000$. The results indicated that the model explained (please check $R^2$) 17% of the variance of voters' attitude toward immigration. The Coefficients' table of Finland's multiple regression analysis is presented in Table 7.7.[14]

Examining the significant beta coefficients in Table 7.7, it is seen that when all the independent variables are put into the regression model, voters' attitude toward immigration can be predicted by their positions within the left-right political spectrum, by their ideas about their country's situation (Finland's general situation), and whether they vote because of immigration or not. Thus, after controlling for the other variables in the model, there will be an expected change in the value of voters' attitude toward immigrants for voting because of immigration or for a one-unit increase in voters' position in the left-right ideological dimension (moving to the right) or in their ideas about Finland's general situation (moving from very good to very bad). Thus, issue voting such as immigration, placing oneself on the right of the political dimension, and having negative ideas about Finland's general situation affect voters' attitude toward migration negatively and would increase the likelihood of having negative attitudes toward immigration.

As expected and discussed in previous sections, Finnish voters, who vote because of the issue of immigration, are expected to be more "alert" about the issue of migration and have a more negative attitude toward migration. The reason might be the anti-immigration discussions that have taken place since 2008. From 2008, after the municipal elections, "anti-immigration issues became the focus of broad interest in media and politics" (Keskinen et al., 2009, in Keskinen, 2013, p. 227) and some of the voters could be influenced by these discussions. The results also show that if voters place themselves on the right (of the left-right political spectrum) then they are expected to have a more negative attitude toward international migration. This might be because they share more conservative thoughts and ideas than the ones who place themselves on the left. Lastly, according to the results, if Finnish voters perceive Finland's general situation, such as economic situations, as poor or worsening, then they are expected to have a more negative attitude toward international migration. This might be because they accuse

*Table 7.8* *Multiple regression analysis for the Hungarian case*

***Hungary-Coefficients***

| | ***Model*** | | Unstandardized Coefficients | | Standardized Coefficients | Sig. |
|---|---|---|---|---|---|---|
| | | | B | Std. Error | Beta | |
| | *1* | **(Constant)** | 1.056 | .203 | | .000 |
| | | Life-Satisfaction (DK out)-recoded | -.023 | .050 | -.021 | .647 |
| | | **The situation in (our) country general-recoded** | .250 | .046 | .268 | .000 |
| | | Voters' internet usage (15 days) before EP elections-recoded | -.044 | .017 | -.111 | .010 |
| | | **Left-Right Political Dimension (5-scale)-recoded** | .150 | .025 | .261 | .000 |
| | | Age (Recoded 4 Categories) | .098 | .035 | .119 | .005 |
| | | **Political Interest Index (D71 SUMMARIZED)** | .125 | .035 | .137 | .000 |
| | | **Reason to Vote (issue cause to vote) – Immigration** | .403 | .073 | .217 | .000 |

*Note:* Dependent variable: average attitude scale toward immigration.

immigrants of worsening the country's general situation or they worry about not having enough resources for all (migrants and "them").

For the Hungarian case, the multiple regression model with all ten predictors produces an adjusted $R^2 = .172$, $F(7, 560) = 17.767$, $p < .000$. This means that 17% of the variance of voters' attitude toward immigration can be predicted from the model. The Coefficients' table of Hungary's multiple regression analysis is presented in Table 7.8.

Table 7.8 shows that Hungarian voters' attitude toward immigration can be predicted by their positions in the left-right political spectrum, their ideas about Hungary's general situation, whether they vote because of immigration or not, and their interest in politics (having political discussions). The independent variables have significant positive regression weights, indicating that, similar to the Finnish case, issue voting as immigration, placing oneself on the right of the political spectrum, having negative ideas about Hungary's general situation affect voters' attitude toward migration negatively and would increase the likelihood of having negative attitudes toward immigration.

Similar to Finnish voters, Hungarian voters who vote because of the issue of immigration are expected to be more "alert" about the issue of migration and have a more negative attitude toward migration. One of the reasons this might be the case in Hungary is because Muslim migrants were "exploited by the governing populist radical right party Fidesz and the right-wing extremist party Jobbik to re-narrate the enemies of the Hungarian nation" (Thorleifsson, 2017, p. 319). It is possible for voters to be influenced by these exploitations. The results also show that if voters place themselves on the right (of the left-right political spectrum) then they are expected to have a more negative attitude toward international migration. Similar to the Finnish case, this might be because they share more conservative thoughts and ideas than those who place themselves on the left. Lastly, if Hungarian voters perceive the country's general situation, such as the economic situation, as poor or worsening, then they are expected to have a more negative attitude toward international migration. As in the Finnish case, this might be because they accuse immigrants of worsening the country's general situation or they worry about not having enough resources for all (migrants and "them"). Different from the Finnish case, the lower the political discussion among the voters, the higher the likelihood of having negative attitudes toward immigration.

*Table 7.9* *Multiple regression analysis for the Bulgarian case*

| ***Bulgaria-Coefficients*** | | | | | | |
|---|---|---|---|---|---|---|
| | ***Model*** | | Unstandardized Coefficients | | Standardized Coefficients | Sig. |
| | | | B | Std. Error | Beta | |
| | ***1*** | **(Constant)** | 1.988 | .197 | | .000 |
| | | Life-Satisfaction (DK out)-recoded | .048 | .039 | .066 | .221 |
| | | **The situation in (our) country general-recoded** | .258 | .043 | .323 | .000 |
| | | Online Social Network Usage (to what extent)-recoded | -.004 | .018 | -.011 | .845 |
| | | Left-Right Political Dimension (5-scale)-recoded | -.038 | .022 | -.088 | .078 |
| | | Age (Recoded 4 Categories) | -.010 | .036 | -.014 | .793 |
| | | Political Interest Index (D71 SUMMARIZED) | .007 | .035 | .008 | .847 |
| | | Reason to Vote (issue cause to vote) – Immigration | .194 | .122 | .068 | .111 |

*Note:* Dependent variable: average attitude scale toward immigration.

For the Bulgarian case, the multiple regression model (simultaneous regression) with all ten predictors produces an adjusted $R^2 = .149$, $F(7, 466) = 12.845, p < .000$, meaning that 15% of the variance of voters' attitude toward immigration can be predicted from the life satisfaction of voters, their ideas about their country's situation in general, their internet usage (15 days) before EP elections, the left-right political spectrum, their interest in politics (having political discussions), the reason they vote (because of immigration), and some demographic characteristics. The Coefficients' table of Bulgaria's multiple regression analysis is presented in Table 7.9.

Examining the significant beta coefficients in the table, it is seen that Bulgarian voters' attitude toward immigration can only be predicted from their ideas about Bulgaria's situation in general. Thus, after controlling for the other variables in the model, there will be an expected change in the value of voters' attitude toward immigration (from positive to negative) for a one-unit increase in voters' ideas about Bulgaria's general situation (moving from very good to very bad). Thus, for the voters, the more they believe that Bulgaria's general situation is bad, then they are expected to have a more negative attitude toward migration (because of, maybe, accusing the immigrants about the situation).

## DISCUSSION AND CONCLUSIONS

As a result of a significant project, the Digital Refuge, it is seen that a number of refugees using social media do not talk about Finland, Bulgaria or Hungary positively when they share posts on social media platforms such as Facebook. Some of the reasons can be a non-welcoming society, politics, policies and/or social context that is reinforced, for instance, through media and the internet. By focusing on these countries (mainly the public), and by comparing Finnish, Bulgarian and Hungarian voters' attitudes toward migration, this study aims to understand whether demographic characteristics, life satisfaction, factors related to social media (such as internet usage frequency), and factors related to voting behavior (such as political interest) have any effect on voters' negative attitudes. The empirical findings suggest that Finnish, Bulgarian and Hungarian voters' attitude toward migration is shaped by their ideas about these countries' general situation. More importantly, issue voting such as immigration, placing oneself on the right of the political spectrum, and low political interest are also important predictors for voters' negative attitude toward migration in some of the cases.

As a result of statistical analyses, it is seen that among demographic variables, the voters' self-placement on the left-right political spectrum has effects on their attitudes toward migration. Regards the factors related to the political behavior of Finnish, Hungarian and Bulgarian voters, the hypothesis tests show that different party supporters in Finland, Bulgaria and Hungary have different attitudes toward immigration (some of them are more negative while some of them are positive toward immigration).

Secondly, the correlation analyses indicate that age and internet usage of voters, their places on the left-right political spectrum and their political interest is related to their attitudes toward migration: for Finland and Hungary, more voters place themselves on the right and the more their interest in politics decreases, the more their attitudes toward migration become negative. In addition, for Hungary and Bulgaria, the more voters' internet usage increases, the more their attitudes toward migration become negative. For Bulgaria, as voters become older and when they place themselves on the right of the political spectrum, their attitude becomes more negative on the issue of migration.

Focusing on the regression results to understand the reasons for the negative attitude toward migration in Finland, Hungary and Bulgaria and to make several predictions, it is claimed that in all three cases, voters' ideas about their countries' general situation is a predictor of their attitude toward migration: having negative ideas about Finland's, Hungary's or Bulgaria's general situation affects voters' attitude toward migration negatively and would increase the possibility of having negative attitudes toward immigration. Thus, if voters argue that their country's situation is bad in general, then it is possible for them to become more negative toward international migration. For the Finnish and Hungarian cases, issue voting such as immigration and placing oneself on the right of the political spectrum affect voters' attitude toward migration negatively and would increase the possibility of having negative attitudes toward immigration. Only for the Hungarian case, if political discussion among the voters is low, is there a higher possibility of voters having negative attitudes toward immigration.

For all of the cases, around 20% of the variance of voters' attitude toward immigration can be predicted from the models. This is a very high percentage for the social sciences; nevertheless, further analysis could be made by using different variables and data sets as well as by using qualitative data.

To conclude, the statistical analyses show that ideas about a country's general situation is a common factor that can explain the level of negative attitude toward immigration in all the three cases. Other reasons are related to individuals' political behavior and their left-right political placements. One policy implication of these findings could be explaining to voters several possible benefits of international migration for their countries' general situation (such as human capital) more clearly in order to increase their level of positive attitudes toward immigration. Of course, as the results show, policy makers and politicians should not "abuse" the subject. Finally, conducting an analysis of the causes and consequences of migration to Finland, Bulgaria and Hungary could be helpful in developing a comprehensive policy framework that will support positive attitudes toward migration.

## NOTES

1. "Adults surveyed in Gallup's World Poll were asked two questions about immigration: 1) In your view, should immigration in this country be kept at its present level, increased or decreased? 2) Do you think immigrants mostly take jobs that citizens in this country do not want (e.g. low-paying or not prestigious jobs), or mostly take jobs that citizens in this country want?" (Esipova et al. 2015, p. 1).
2. The official final turnout at the European level was announced as 50.66% and the voting turnout of these countries are 40.8%, 43.36% and 32.64% respectively. See https://www.europarl.europa.eu/news/en/press-room/20191029IPR65301/final-turnout-data-for-2019-european-elections-announced (accessed August 24, 2020).
3. As an example, UK Independence Party voters can be divided into two groups: "'strategic' supporters who only vote UKIP at EP elections and 'core' supporters who also vote UKIP at Westminster elections" (Ford et al., 2012, p. 207). Strategic supporters seem to be "Conservative voters registering their hostility to the EU while core supporters are a poorer, more working class and more deeply discontented group who more closely resemble supporters of the BNP and of European radical right parties (Ford & Goodwin 2010; Mudde 2007)" (in Ford et al., 2012, p. 207). As a result, it can be argued that "UKIP's credentials as a legitimate party of rightwing protest over Europe may make it a 'polite alternative' for voters angry about rising immigration levels or elite corruption but who are repelled by the stigmatised image of the more extreme BNP which, as polling data reveals, has struggled to portray itself as a credible political choice (Goodwin 2010; John & Margetts 2009)" (in Ford et al., 2012, p. 207).

4. Hungary (number of posts: 71 and average sentiment score: 1.68), Finland (number of posts: 44 and average sentiment score: 1.77) and Bulgaria (number of posts: 28 and average sentiment score: 1.65). See https://digitalrefuge.berkeley.edu (accessed August 24, 2020).
5. In Standard Eurobarometer 91 Spring 2019 Public opinion in the European Union First results report, the sampling design is explained as: "The basic sample design applied in all states is a multi-stage, random (probability) one. In each country, a number of sampling points was drawn with probability proportional to population size (for a total coverage of the country) and to population density. In order to do so, the sampling points were drawn systematically from each of the 'administrative regional units', after stratification by individual unit and type of area. They thus represent the whole territory of the countries surveyed according to the EUROSTAT NUTS II (or equivalent) and according to the distribution of the resident population of the respective nationalities in terms of metropolitan, urban and rural areas. In each of the selected sampling points, a starting address was drawn, at random. Further addresses (every Nth address) were selected by standard 'random route' procedures, from the initial address. In each household, the respondent was drawn, at random (following the 'closest birthday rule')."
   Hungary 95% confidence level with its population size (+15) 8.781.161; Finland 95% confidence level with its population size (+15) 4.747.810; and Bulgaria 95% confidence level with its population size (+15) 6.537.535.
6. Exploratory factor analysis (EFA) was conducted using Principle Components method followed by Direct Oblimin rotation. Oblimin rotation was used because selected variables were expected to be correlated. Kaiser-Meyer-Olkin Measure of Sampling Adequacy Test (KMO test) value is 0.717 indicating that the sampling is adequate. Similarly, Bartlett's test of sphericity with its small value (0.000) of significance level indicates that a factor analysis may be useful with the data. EFA suggests single factor solution. This factor accounts for 55.1% of the total variance. The reliability analysis (Cronbach's alpha is 0.725) indicates a high level of internal consistency for these variables with the specific sample.
7. For Finland, $t_{552.377} = 2.117, p > .01$, for Hungary $t_{585.813} = -.520, p > .01$, and for Bulgaria $t_{523.501} = -.884, p > .01$.
8. $(F\ 11{,}554) = 8.493, p = .000$.
9. For Hungarian case, $\chi 2(2) = 41.432, p = .000$.
10. For Bulgarian case, $\chi 2(2) = 43.092, p = .000$.
11. It is based on 560 Finnish participants with 2-tailed significance, $p = .001$ meaning that there is a .001 probability of finding this sample correlation, if the real population correlation is zero.
12. It is based on 582 Hungarian participants with 2-tailed significance, $p = .001$.
13. It is based on 559 Bulgarian participants with 2-tailed significance, $p = .001$ meaning that there is a .001 probability of finding this sample correlation, if the real population correlation is zero.

14. The outputs of the cases show that the VIF for variables in analysis are between 1.05 and 1.8 indicating some correlation, but not enough to be overly concerned about multicollinearity.

# BIBLIOGRAPHY


2019 European Election Results, accessed August 21, 2020 at https://www.europarl.europa.eu/election-results-2019/en

Albrecht, S. (2006), "Whose voice is heard in online deliberation?: a study of participation and representation in political debates on the internet", *Information, Communication & Society* 9(1), 62–82.

Ardèvol-Abreu, A. and Gil de Zúñiga, H. (2017), "Effects of editorial media bias perception and media trust on the use of traditional, citizen, and social media news", *Journalism and Mass Communication Quarterly* 94(3), 703–24.

Bartlett, J. (2014), "Populism, social media and democratic strain", in C. Sandelind (ed.), *European Populism and Winning the Immigration Debate*, ScandBook, Falun: FORES, ELF.

Bobeva, D. (2017), *Migration: Recent Developments in Bulgaria.* The report at SSRN: https://ssrn.com/abstract=2929544

Dalton, R. (2000), "Citizen attitudes and political behavior", *Comparative Political Studies* 33(6/7), 912–40.

Esipova, N., Ray, J., Pugliese, A. and Tsabutashvili, D. (2015), *How the World Views Migration*, Switzerland: International Organization for Migration.

Eurobarometer 91.5 Data (Media Monitoring and Eurobarometer) (2019), accessed May 29, 2020 at https://search.gesis.org/research_data/ZA7576

Flores, R.D. (2018), "Can elites shape public attitudes toward immigrants?: evidence from the 2016 US Presidential Election", *Social Forces 96*, 1649–90.

Ford, R., Goodwin, M.J. and Cutts, D. (2012), "Strategic Eurosceptics and polite xenophobes: support for the United Kingdom Independence Party (UKIP) in the 2009 European Parliament", *European Journal of Political Research* 51, 204–34.

Hansen, R. (2003), "Migration to Europe since 1945: its history and its lessons", *Political Quarterly* 74(1), 25–38.

Jasinskaja-Lahti, I., Liebkind, K., Horenczyk, G. and Schmitz, P. (2003), "The interactive nature of acculturation: perceived discrimination, acculturation attitudes and stress among young ethnic repatriates in Finland, Israel and Germany", *International Journal of Intercultural Relations* 27, 79–97.

Kalaycioğlu, E. (1983), *Karşılaştırmalı Siyasal Katılma* [Comparative Political Participation], İstanbul: İstanbul Üniversitesi Siyasal Bilimler Fakültesi Yayınları.

Keskinen, S. (2013), "Antifeminism and white identity politics: political antagonisms in radical right-wing populist and anti-immigration rhetoric in Finland", *Nordic Journal of Migration Research* 3(4), 225–32.

Leeper, T. and Slothuus, R. (2014), "Political parties, motivated reasoning, and public opinion formation", *Political Psychology* 35, 129–56.

Mayda, A. (2006), "Who is against immigration? A cross-country investigation of individual attitudes toward immigrants", *The Review of Economics and Statistics* 88(3), 510–30.
Neiman, M., Johnson, M. and Bowler, S. (2006), "Partisanship and views about immigration in Southern California: just how partisan is the issue of immigration?", *International Migration* 44(2), 35–56.
Nonchev, A., Encheva, S. and Atanassov, A. (2012), "Social inequalities and variations in tolerance toward immigrants", *International Journal of Sociology* 42(3), 77–103.
O'Rourke, K.H. and Sinnott, R. (2006), "The determinants of individual attitudes towards immigration", *European Journal of Political Economy* 22, 838–61.
Papacharissi, Z. (2002), "The virtual sphere: the internet as a public sphere", *New Media & Society* 4(1), 9–27.
Press Releases, Final turnout data for 2019 European elections announced (2019), News, accessed August 24, 2020 at https://www.europarl.europa.eu/news/en/press-room/20191029IPR65301/final-turnout-data-for-2019-european-elections-announced
Press Releases, 2019 EP Election, News, accessed August 21, 2020 at https://www.europarl.europa.eu/news/en/press-room/20190923IPR61602/2019-european-elections-record-turnout-driven-by-young-people
Raycheva, L. and Peicheva, D. (2017), "Populism in Bulgaria between politicization of media and mediatization of politics", *Mediatization Studies* 1, 69–81.
Standard Eurobarometer 91 (Spring 2019), "Public opinion in the European Union First results", European Union.
Stefanova, B. (2009), "Ethnic nationalism, social structure, and political agency: explaining electoral support for the radical right in Bulgaria", *Ethnic and Racial Studies* 32(9), 1534–56.
Szoke, L. (1992), "Hungarian perspectives on emigration and immigration in the new European architecture", *The International Migration Review* 26(2), 305–23.
The Digital Refuge Project, Exploring the European Refugee Crisis (2018), accessed May 29, 2020 at https://digitalrefuge.berkeley.edu
Thorleifsson, C. (2017), "Disposable strangers: far-right securitisation of forced migration in Hungary", *Social Anthropology/Anthropologie Sociale* 25(3), 318–34.
Van Der Brug, W., Fennema, M. and Tillie, J. (2000), "Anti-immigrant parties in Europe: ideological or protest vote?", *European Journal of Political Research* 37, 77–102.
Wallis, E. (2019), "What the European Parliament election results could mean for migration", accessed August 21, 2020 at https://www.infomigrants.net/en/post/17132/what-the-european-parliament-election-results-could-mean-for-migration
Ward, C. and Masgoret, A. (2008), "Attitudes toward immigrants, immigration, and multiculturalism in New Zealand: a social psychological analysis", *The International Migration Review* 42(1), 227–48.

Weeks, B.E., Lane, D.S., Kim, D.H., Lee, S.S. and Kwak, N. (2017), "Incidental exposure, selective exposure, and political information sharing: integrating online exposure patterns and expression on social media", *Journal of Computer-Mediated Communication* 22(6), 363–79.

# Index